Sequential Simplex Optimization

Sequential Simplex Optimization

A Technique for Improving Quality and Productivity
in Research, Development, and Manufacturing

Frederick H. Walters
Lloyd R. Parker, Jr.
Stephen L. Morgan
Stanley N. Deming

CRC Press, Inc.
Boca Raton, Florida

Library of Congress Cataloging-in-Publication Data

Sequential simplex optimization : a technique for improving quality
    and productivity in research, development, and manufacturing /
    authors, Frederick H. Walters ... [et al.].
        p.   cm. - (Chemometrics series)
    Includes bibliographical references and index.
    ISBN 0-8493-5894-9 : $49.95
        1.   Chemical process control.   2.   Operations research.
    3.   Factorial experiment designs.   I.   Walters, Frederick H.
    II.   Series: Chemometrics series (Boca Raton, Fla.)
    TP155.75.S45   1991
    660'.2815-dc20   91-14187
                                                                                 CIP

Direct all inquiries to CRC Press, Inc., 2000 Corporate Blvd., N.W.,
Boca Raton, Florida 33431.

© 1991 by CRC Press, Inc.

International Standard Book Number 0-8493-5894-9

Library of Congress Card Number 91-14187
Printed in the United States of America   3 4 5 6 7 8 9 0

Printed on acid-free paper

Chemometrics Series

Steven D. Brown, Editor
University of Delaware

Advisory Editors

Lloyd A. Currie
National Institute of Standards and Technology

Paul Geladi
University of Umeå

Paul J. Gemperline
East Carolina University

Phillip K. Hopke
Clarkson University

The objective of this series is to publish a broad range of introductory, intermediate, and research level texts on all aspects of mathematical and statistical analysis of chemical data. Topics of interest for the series include applied statistics, optimization, factor analysis, calibration, pattern recognition, chemical modeling and parameter estimation, signal processing, resolution, structure-property relationships, library searching, and artificial intelligence. The theoretical aspects of any of these chemometric methods or the application of these methods are appropriate subjects for books in the series. Chemometrics applications books on closely related fields, such as materials chemistry, pharmacy, and food science are also welcome.

Other Titles

Factor Analysis of Chromatographic Data, P. Gemperline and S. Ramos

Applications of Numerical Methods in Molecular Spectroscopy, P. Pelikán, M. Čeppan, and M. Liška

Series Preface

At a time when the computer has become so commonplace and powerful, it is surprising that few sources of useful information exist for the chemist who seeks to take advantage of this cheap computing power for the analysis of experimental data and the design and optimization of experiments. Getting the most from the newer methods for design of experiments and analysis of data – either with the aid of commercial software packages or with user-developed code – requires much more than just easy access to a computer. A good understanding of theory and experience with practical analysis can make the difference between meaningful results and meaningless sets of numbers. This series has as a goal the publishing of texts that will make the conversion of numbers to results a little easier for students and practicing chemists. Those with an interest in the collection and analysis of data from chemical systems will find covered in this series both the theory and practical applications needed for the efficient use of the computing power available on their desks.

The texts in this series will be useful not only to students of chemistry and practicing chemists, but also to those in related fields, such as food science, pharmacy, materials science and environmental science, where the analysis of chemical measurements plays a central role.

Steven D. Brown
Newark, Delaware

Preface

The sequential simplex methods described in this text are *not* the methods used in linear programming (the simplex tableaux). The sequential simplex methods described in this text are *not* the lattice designs used in formulations work (simplex mixture designs). The sequential simplex methods described in this text *are* methods used in evolutionary operation (EVOP). All three methods (sequential, linear programming, and formulations) make use of a geometric figure called a simplex, but each method uses the simplex in a distinctive way.

We like the sequential simplex methods. They are easy to use and can be used to rapidly optimize systems containing several continuous factors. If their limitations are recognized, the sequential simplex methods can be used successfully by almost anyone in almost any field. But we are chemists, and we would like to tell you how we became interested in the sequential simplexes.

In the mid-1960s, chemists were discovering the wonders of computers and computerized automation. Previously developed routine operations that had required massive amounts of time and human energy could now be carried out rapidly and autonomously. It became possible to robotize many existing chemical operations.

Toward the late 1960s, some chemists were beginning to explore the possibility of using computers and computerized automation for a different purpose. They were interested in automatically carrying out the *nonroutine* operations that are required to develop a chemical method. Any chemical method requires a "recipe"–the specification of values for a large number of variables, such as reagent concentrations, temperatures, and times. The recipe is developed by experimentation. In the late 1960s, chemists were wondering if computers and automation could be used to carry out the nonroutine experiments required to arrive at these specifications.

As it turned out, it was a relatively easy task to assemble the pieces of automated apparatus and control them with computers. Automated chemical experiments became a reality. But the reality forced chemists to confront directly a question that had until then remained in the background: "What kinds of experiments should be done to develop a chemical method?"

A bit of philosophical thought suggested that three basic tasks were probably required to develop a chemical method: (1) obtain some response, (2) improve that response, and (3) understand that response.

The first task could be accomplished with chemical knowledge – a knowledge of chemical reactions, of the interaction of physical phenomena with matter, etc. – the kinds of things taught to chemists in undergraduate and graduate programs. But the second and third tasks seemed to fall outside chemistry. Unfortunately for chemists, the second and third tasks seemed to lie in that dreaded area known as "statistics."

In fact, the subdiscipline of statistics known as "experimental design" is just the thing for improving and understanding many types of chemical systems. Chemists in England and elsewhere had been using classical factorial designs to improve the yield of large-scale chemical reactions since the mid-1940s. One technique was known as "evolutionary operation," or EVOP, for short. The improvement of yield and, along with it, the improvement of fundamental chemical understanding that arose from the results of these experiments were unprecedented in the history of chemistry.

In 1962, W. Spendley, G. R. Hext, and F. R. Himsworth at Imperial Chemical Industries Ltd. in England published a paper in which they set out, among other things, to make EVOP automatic ["Sequential application of simplex designs in optimisation and evolutionary operation," *Technometrics*, 4, 441–461 (1962)]: "The work now to be described sprang originally from consideration of how Evolutionary Operation might be made automatic – a possibility to which [G. E. P.] Box drew attention even in 1955. More specifically, two problems were posed, viz. (i) could an Evolutionary Operation procedure be devised which would more rapidly approach and attain optimum conditions, and (ii) *could the calculations and decisions be so formalised and simplified that they could be executed automatically by a digital computer*" [emphasis ours]. That last problem was clearly of interest to chemists who wanted to use computers and computerized automation to develop new chemical methods. And so we became interested in sequential simplex optimization.

Although Spendley, Hext, and Himsworth were looking toward the day when computers could use the sequential simplex, the calculations themselves are very simple and can be done by hand on a worksheet. A computer is not required for the successful use of sequential simplex methods.

The material in this book grew out of a short course offered for the first time in the summer of 1975 at the University of Houston. It has been presented more than 80 times since then. We do not believe that a course on optimization must necessarily be preceded by a course on statistics. Instead, we have taken the approach that both subjects can be developed simultaneously, complementing each other as needed, in a course that presents the fundamentals of the sequential simplex method.

It is our intent that the book can be used in a number of fields by advanced undergraduate students, by beginning graduate students, and (perhaps more important) by workers who have already completed their formal education. The material in this book has been presented to all three groups, either as part of regular one-semester courses, or through intensive two- or three-day short courses. We have been pleased by the confidence these students have gained from the courses, and by their enthusiasm as they apply these methods in their various disciplines.

The core of the book is contained in Chapters 3 and 4, in which the fixed-size algorithm of Spendley, Hext, and Himsworth, and the variable-size algorithm of Nelder and Mead are presented. Chapter 5 compares the behavior of the fixed- and variable-size simplexes. Chapters 6 and 7 contain additional information generally needed for the successful application of the sequential simplex.

Chapter 2 attempts to present some of the history behind the development of the sequential simplex method and to show how it grew out of the practice of factorial EVOP. Chapter 1 attempts to present what we believe will be the future of the sequential simplex method—a tool for achieving quality and competitiveness.

Chapter 8 is devoted to desirability functions for multicriteria decision making. Because the sequential simplex method *is* a sequential method, trade-offs among multiple responses must be made as the simplex progresses. Desirability functions are one means whereby these multiple responses can be combined into a single figure-of-merit to drive the simplex. Chapter 9 presents a brief introduction to linear models and classical response surface designs for use after the simplex has brought the system into the region of the optimum.

Chapter 10 reviews representative applications of the sequential simplex in chemistry and related fields. Chapter 11 is a more complete bibliography of publications using the sequential simplex methods. Most of these papers are applications (experimental optimization of systems in the real world), but interesting theoretical papers are also listed.

We are grateful to a number of friends for help in many ways. Grant Wernimont (now deceased) introduced SND to Box-type EVOP while SND was a graduate student and Grant was a visiting professor at Purdue in the late 1960s. Grant sowed the seed that eventually got all of us authors interested in optimization and experimental design. His many years of industrial experience at Kodak gave an applied perspective to subjects that could otherwise become highly academic. We thank L. B. Rogers for his insistence that analytical chemists should learn about statistics and experimental design; he was right (as usual!). Paul G. King deserves special recognition for his willingness to discuss the fine points of simplex optimization. His suggestion of a modification of the Nelder and Mead algorithm to give it even greater efficiency in the real world is a reflection of his usual brilliant thought.

W. Spendley provided personal insight about the development of the sequential simplex methods; we are grateful for his comments. Wendy Aldwyn skillfully translated a difficult concept into Figures 7.34 and 7.35. We thank David Councilman for careful typing of the manuscript. David and Sandra Councilman and Robin and Jim Corbit read the page proofs; any remaining errors are the responsibility of the authors. One of us (SND) acknowledges the University of Houston Faculty Development Leave Program for time to bring this project to completion.

Finally, we would like to acknowledge our students who provided criticism as we developed the material presented here.

FREDERICK H. WALTERS
Lafayette, Louisiana

LLOYD R. PARKER, JR.
Oxford, Georgia

STEPHEN L. MORGAN
Columbia, South Carolina

STANLEY N. DEMING
Houston, Texas

Frederick H. Walters received his Bachelor of Science degree from the University of Waterloo (Waterloo, Ontario, Canada) in 1971. He received his Ph.D. from the University of Massachusetts (Amherst, Massachusetts) in 1976 working under Professor Peter C. Uden. He is presently an Associate Professor of Chemistry at the University of Southwestern Louisiana (Lafayette, Louisiana).

Lloyd R. Parker, Jr. received his Bachelor of Arts degree from Berry College (Mount Berry, Georgia) in 1972. He received his Master of Science degree from Emory University (Atlanta, Georgia) in 1974 and his Ph.D. from the University of Houston (Houston, Texas) in 1978 working under Professor Stanley N. Deming. He is presently an Associate Professor of Chemistry at Oxford College of Emory University (Oxford, Georgia).

Stephen L. Morgan received his Bachelor of Science degree from Duke University (Durham, North Carolina) in 1971. He received his Master of Science degree (1974) and his Ph.D. (1975) from Emory University (Atlanta, Georgia), working under Professor Stanley N. Deming. He is presently a Professor of Chemistry at the University of South Carolina (Columbia, South Carolina).

Stanley N. Deming received his Bachelor of Arts degree from Carleton College (Northfield, Minnesota) in 1966. He received his Master of Science and Ph.D. degrees from Purdue University (Lafayette, Indiana) in 1970 and 1971 working under Professor Harry L. Pardue. He is presently a Professor of Chemistry at the University of Houston (Houston, Texas).

Contents

Chapter **1**

The Need for Optimization*

In 1939, in the introduction to Walter A. Shewhart's famous book, *Statistical Method from the Viewpoint of Quality Control*, W. Edwards Deming wrote [1],

> Most of us have thought of the statistician's work as that of measuring and predicting and planning, but few of us have thought it the statistician's duty to try to bring about changes in the things that he measures. It is evident, however, that this viewpoint is absolutely essential if the statistician and the manufacturer or research worker are to make the most of each other's accomplishments.

A few parts of the American economy have grown exceptionally strong, at least in part because they adopted the statistical philosophies of Walter A. Shewhart, W. Edwards Deming, Joseph M. Juran, and others [2, 3]. Many foreign economies have grown especially strong, largely because of their emphasis on statistical methods. This has resulted in significant foreign competition that now appears to have had adverse effects on the American economy, particularly in the areas of automobile and related manufacturing.

In the Youden Memorial Address at the 1984 ASA-ASQC (American Statistical Association, American Society for Quality Control) Fall Technical Conference, Brian L. Joiner repeated W. Edwards Deming's earlier comment with a new urgency [4]:

*Adapted with permission from Deming, S.N. *CHEMTECH*, September 1988, *18* (9), 560–566. Copyright 1988. American Chemical Society.

> There is much to be done if North American industry is to survive in the new
> economic age. We statisticians have a vital role to play in the transformation
> that is needed to make our industry competitive in the world economy.

After almost half a century, American industry finally seems ready to accept the
statistician's long-standing offer of help.

QUALITY

The focus of the current industrial transformation is "quality," a term that is difficult to define. As Joseph M. Juran says, "There is no known short definition that
results in a real agreement on what is meant by quality" [5]. Juran goes on to say
that quality has several meanings, two of which are critical for both quality planning and strategic business planning:

1. Product performance and product satisfaction. Features such as "promptness of
    a process for filling customer orders," "fuel consumption of an engine," and
    "inherent uniformity of a production process" determine the customers' satis-
    faction with a product. "Because of the competition in the market place, a
    primary goal for product performance is to be equal or superior to the quality
    of competing products" [5].
2. Freedom from deficiencies and product dissatisfaction. Features such as "late
    deliveries," "factory scrap or rework," and "engineering design changes" deter-
    mine the customers' dissatisfaction with a product. "Some deficiencies impact
    external customers and hence are a threat to future sales as well as a source of
    higher costs. Other deficiencies impact internal customers only and hence are
    mainly a source of higher costs.... For quality in the sense of freedom from
    deficiencies, the long range goal is perfection" [5].

It is clear from the above considerations of quality that [4]

> The key to improved quality is improved processes.... Processes make
> things work. Thousands of processes need improvement, including things
> not ordinarily thought of as processes, such as the hiring and training of
> workers. We must study these processes and find out how to improve them.
> The scientific approach, data-based decisions, and teamwork are key to
> improving all of these processes.

STATISTICAL PROCESS CONTROL

"Statistical process control" (SPC) is another term that is difficult to define. One
fundamental explanation is that, "With the help of numbers, or data, [STATISTI-
CAL] [we] study the characteristics of our process [QUALITY] [in] order to make it
behave the way we want it to behave [CONTROL]" [6]. This definition, with emphasis on making the process behave the way we want it to behave, is probably closest
to what is involved in the current transformation of American industry.

Other terms that were originally used to mean the same thing have become
suspect in recent years. "Statistical quality control" (SQC) and "quality assurance"

(QA) could be synonymous with SPC, but many industries have achieved SQC and QA not by making the process produce good product, but rather by mass inspection of the outgoing product – separating the good from the bad. Only in rare situations is mass inspection appropriate. In general, "Inspection is too late, ineffective, costly. When a lot of product leaves the door of a supplier, it is too late to do anything about the quality of the lot. Scrap, downgrading, and rework are not corrective action on the process. Quality comes not from inspection, but from improvement of the process" [7].

Almost all modern quality writers stress that management should fulfill a leadership role in the improvement of quality [8]. The responsibility for quality cannot be delegated by management to others. Quality assurance is not the responsibility of just one group at the bottom of an organizational chart. Quality assurance is the responsibility of everyone within the organization. Management is ultimately responsible for quality. As Berger and Hart emphasize [9],

> If a process is not capable, then before SPC can be implemented, management, *MANAGEMENT*, must take action to correct the process. If management cannot or will not act to correct the process then they can expect no improvement in the quality of output from any quality improvement program, much less from SPC. [Emphasis theirs.]

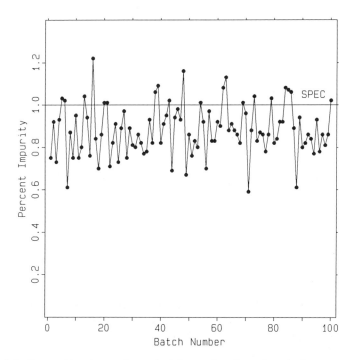

Figure 1.1 Percentage impurity vs. batch number for a chemical process operated near pH 6.3. See text for additional details. The data in this figure were adapted from Problem 2-1 in [11], offset, scaled, and renamed.

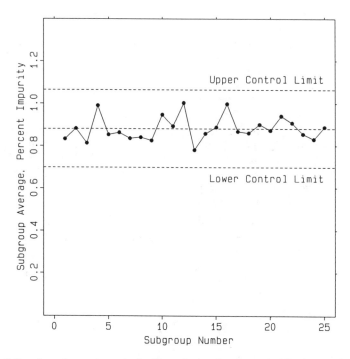

Figure 1.2 *x*-bar (mean) control chart from the data in Figure 1.1. See text for details.

Management should also realize [*10*]

> that SPC [by itself] is not a cure-all for quality and production problems. It will not correct a poor product design or resolve poor employee job training. Nor will it correct an inefficient process or worn out machines and tooling. However, it will help in leading to the discovery of all of these types of problems and to identifying the type and degree of corrective action required.

One SPC technique for discovering some of these problems and monitoring improvement is the control chart.

CONTROL CHARTS

Consider a chemical process (operated at pH 6.3) that produces not only the desired compound in high yield, but also a relatively small amount of an undesirable impurity. Discussions between the producer and consumer of this material suggest that an impurity level up to 1.0% can be tolerated, but an impurity level greater than this is unacceptable. By mutual consent, a specification level of <1.0% is set.

Figure 1.1 plots the percentage impurity vs. batch number for this chemical process [*11*]. Much of the time the percent impurity is less than 1.0%, but about one

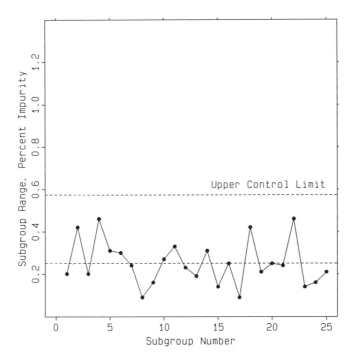

Figure 1.3 r (range) control chart from the data in Figure 1.1. See text for details.

batch in five is outside the specification. This can be costly for the manufacturer if there is no other customer willing to purchase "out-of-spec" material. These out-of-spec batches might be kept in a holding area until they can be reworked, usually by blending with superior grade material. But storage and rework are costly and ultimately weaken the competitive position of the manufacturer.

Figure 1.1 is a way of letting the process talk to us, a way of letting it tell us how it behaves [12]. It seems to be telling us that (1) on the average, the impurity level is below 1.0%, (2) there is some variation from batch to batch, and (3) the process behaves consistently – a moving average would not appear to go up or down very much with time, and the variation seems to be fairly constant with time.

These ideas are confirmed in the statistical "x-bar" and r charts shown in Figures 1.2 and 1.3, respectively. To construct these charts, the group of 100 batches has been subdivided into 25 sequential subgroups of four. The average ($\bar{x}$, pronounced "x-bar") and range (r) have been calculated for each subgroup. The resulting values are plotted as a function of subgroup number. (Subgroups are necessary, in part, to obtain estimates of the range.) As expected, the average percentage impurity does not go up or down very much with time, and the range (a measure of variation) is fairly constant with time.

Making the assumption that the process is stable, so-called "three-sigma" limits based on statistical expectation can be calculated for the subgroup average. In Figure 1.2 these limits are labeled the "Upper Control Limit" and the "Lower Control Limit." The middle, unlabeled dashed line in Figure 1.2 is the "Grand

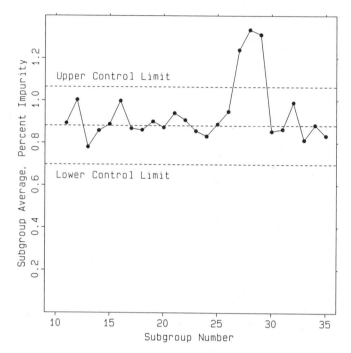

Figure 1.4 Effect of an out-of-control situation on the x-bar control chart.

Average," or the average of averages. With three-sigma limits, the subgroup averages will lie within these limits approximately 99.7% of the time if the process is stable; only very rarely would a subgroup average lie outside these limits if the process is "in statistical control" (that is, if the process is stable).

Similar three-sigma limits can be placed on the range. In Figure 1.3 only the upper control limit is labeled—the lower control limit is zero for this case. The unlabeled dashed line represents the average subgroup range. Only very rarely (approximately 0.3% of the time) would a subgroup range be greater than the upper control limit if the process is in statistical control.

It is absolutely essential to understand that these control limits are a manifestation of the process speaking to us, telling us how it behaves. These control limits do not represent how we would like it to behave. It is a common but misguided practice to draw on control charts lines that represent our wishes. These lines can have no effect on the behavior of the process.

Control charts are useful because they offer a way of letting the process tell us when it has changed its behavior. In Figure 1.4 it is clear that something significant happened at subgroup numbers 27–29. The process has clearly "gone out of control." This many excursions this far away from the control limit in this short a time would be highly unlikely from a statistical point of view if the process were still operating as it was before.

Such excursions suggest that there is some assignable (Walter A. Shewhart) or special (W. Edwards Deming) cause for the observed effect. Because these excursions are undesirable in this example (most of the individual batches produced would probably be unfit for sale), it would be economically important to discover the assignable cause and prevent its occurrence in the future.

It should be noted that excursions in the other direction (below the lower control limit) also suggest the presence of assignable causes. Unfortunately, excursions in this direction often receive less attention. After all, in this example the product can still be sold. The engineer in charge might think, "Isn't that special? I wonder why the process worked so well on those days. Oh well, I can't investigate it now. I have to finish this report on last month's out-of-spec material for management by tomorrow." This is unfortunate, because if the engineer found the assignable cause of this good effect and changed the process accordingly, the engineer would spend less time writing reports on out-of-spec material.

One of the most powerful features of control charts is their ability to show when special causes are at work so that steps can be taken to discover the identity of these special causes and use them to improve the process. However, there are two difficulties with this use of control charts.

First, we have to wait for the process to speak to us. This represents an inefficient, passive approach to process optimization. If the process always stays in statistical control, we will not learn anything and the process cannot get better. In the example of Figure 1.1, we will continue to produce batches of off-spec material about 20% of the time. It is not enough just to be in statistical control—our product must become "equal or superior to the quality of competing products" [5]. We probably cannot wait for the process to speak to us. We must take action now.

Second, when the process does speak to us (when it goes out of control), it tantalizes us by saying, "Hey! I'm behaving differently now. Try to find out why." Discovering the reason the process behaves differently requires that we determine the cause of a given effect. Discovering which of many possible causes is responsible for the observed effect is often incredibly difficult. It is an activity that continues to puzzle philosophers.

EXPERIMENTAL DESIGN

Paul W. Holland, in a paper on cause and effect relationships, concludes that [13]

> The analysis of causation should begin with studying the effects of causes rather than the traditional approach of trying to define what the cause of a given effect is.

That is a powerful conclusion. It is a recommendation of experimental design, whereby the researcher intentionally produces causes (changes in the way the process is operated) and sees what effects are produced (changes in the way the process behaves). With designed experiments, information can be obtained now. We can make the process talk to us. We can ask the process questions and get answers from it. We do not have to wait.

Management sometimes (unknowingly?) uses designed experiments to discover the effects of causes, especially in the evaluation of new employees. "Let's put Jane in charge of the quality control lab for a year. That way we'll learn a lot about her managerial skills so we can decide if she's capable of a bigger assignment." Or a favorite of the petroleum industry, "Send Richard up the chain of command with

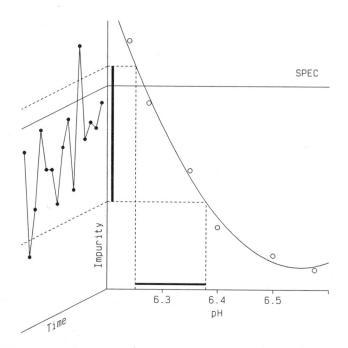

Figure 1.5 Results of a set of experiments designed to determine the influence of pH on percentage impurity. See text for details.

a request for the Dry Hole Production Report. Then we'll see how good he is at meeting people and reacting to criticism."

This same approach can be used with chemical processes as well. "Let's vary the pH of our process and see what effect it has on the impurity level. With that knowledge, we might be able to decrease the impurity level by adjusting the pH." Figure 1.5 contains the results of a set of experiments (open circles) designed to discover the effect of pH on impurity. This represents a scientific approach; it allows us to make data-based decisions [4].

The right side of Figure 1.5 shows the presumed "cause and effect" relationship between percentage impurity and pH that is implied by the results of the experimental design. From the shape of the fitted curve, it would appear that our current operating pH of 6.3 is not optimal. But there are two reasons it is not optimal: not only is the level of impurity relatively high there, but the amount of variation is also relatively high there.

"Set-point control" is almost never set-point control. We might set the controller to maintain a pH of 6.3000, but time constants within the control loop and variations in mixing, temperature, flow rates, sensors, etc. prevent the controller from maintaining a pH of exactly 6.3000. In practice, the pH of the process will fluctuate around the set point. This variation in pH is represented by the black horizontal bar along the pH axis in Figure 1.5.

Variations in pH will be transformed by the process into variations in percentage impurity. The relationship between percentage impurity and pH is rather

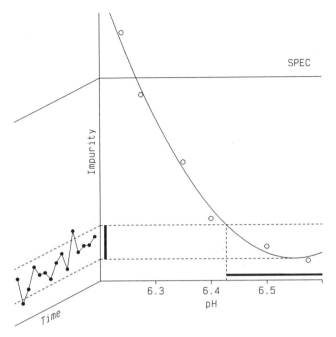

Figure 1.6 Improved control limits for process operated near pH 6.5. See text for details.

steep in the region of pH 6.3. When the pH wanders to lower levels, the percent impurity will be high. When the pH wanders to higher levels, the percentage impurity will be low. This variation in impurity is represented by the black vertical bar along the impurity axis in Figure 1.5.

The left side of Figure 1.5 suggests that variations in pH will, over time, be transformed into variations in percentage impurity and will result in a runs chart similar to that shown in Figure 1.1 when the process is operated in the region of pH 6.3.

How could we decrease the variation in percentage impurity? One way would be to continue to operate at pH 6.3 but use a better, more expensive controller that allows less variation in the pH. This would decrease the width of the "black horizontal bar" in Figure 1.5 (variation in pH), which would be transformed into a shorter "black vertical bar" (variation in impurity). The resulting runs chart would show less variation. But this is a "brute force" way of decreasing the variation. There is another way to decrease the variation in this process.

Other effects being equal, it is clear from the right side of Figure 1.5 that we should change our process's operating conditions to a pH of about 6.5 if we want to decrease the level of impurity. In this example, there is an added benefit from working at these conditions. This benefit is shown in Figure 1.6.

Not only has the level of impurity been reduced, but the variation in impurity has also been reduced. This is because the relationship between impurity and pH is not as steep in the region of pH 6.5. When the process is operated in this region,

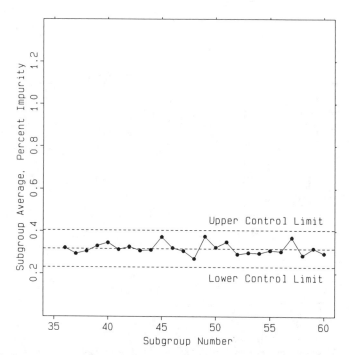

Figure 1.7 *x*-bar control chart for the process operated near pH 6.5. See text for details.

it is said to be "rugged" or "robust" with respect to changes in pH – the process is relatively insensitive to small changes in pH [*14*]. (In this example, the relationship is so flat in the region of pH 6.5 that we could actually allow greater variation in pH and still obtain less variation in percentage impurity.) This principle of ruggedness is one aspect of the Taguchi philosophy of quality improvement [*15*].

If the process were operated at this new pH, the corresponding control chart would be similar to that shown in Figure 1.7, and the corresponding runs chart would look like Figure 1.8.

Some people criticize runs charts like Figure 1.8 as belonging to "gold-plated processes," that are "better than they need to be." In some cases the criticism might be justified – for example, if the economic consequences of operating at the higher pH were not justified by the economic consequences of producing such a good product. But in many cases there are three arguments that speak in favor of these "gold-plated processes."

First, engineers and scientists do not have to spend as much time writing "out-of-spec" reports for management. And, of course, management does not have to spend as much time with customer complaints. No one wastes time on nonproductive "fire-fighting." Human effort can be used in better, more productive ways. All of this because the process behaves the way we want it to behave.

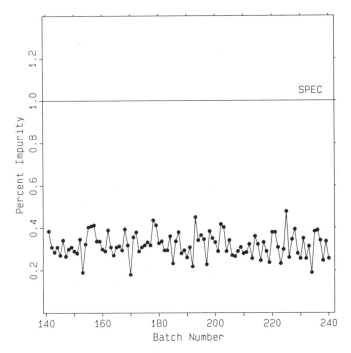

Figure 1.8 Percentage impurity vs. batch number for a chemical process operated near pH 6.5. See text for details.

Second, improvement of product often opens up new markets. Another customer might have an application for our product where 1.0% impurity would be unacceptable, but 0.5% could be tolerated.

Third, there is a lot of "elbow room" between the percentage impurity produced by the improved process and the original specification limit in Figure 1.8. If the process starts to drift upward (perhaps a heat exchanger elsewhere in the process is fouling and this causes the percentage impurity to increase), within-spec material can still be produced while the special cause is discovered and eliminated.

QUALITY BY DESIGN

We have seen how experimental design can be used to improve some of the quality characteristics of a process that is already in production. But should it be necessary to do this? Is manufacturing the group that is responsible for quality? What about the quality of the processes that will be used five years from now? Who is responsible for the quality of those new processes? Who should have been responsible for the quality of existing processes when they were new? Where does quality begin?

Consideration of quality in manufacturing should begin before manufacturing starts to produce the product. This is known as "quality by design" [16]. Just as

Figure 1.9 Producer–consumer relationships between R&D and manufacturing, and between manufacturing and the customer.

there is a producer–consumer relationship between manufacturing and the customer, so too there is a producer–consumer relationship between R&D and manufacturing (Figure 1.9). Manufacturing (the consumer) should receive from R&D (the producer) a process that has good quality characteristics. In particular, R&D should develop, in collaboration with manufacturing, a process that is rugged with respect to anticipated manufacturing variables [17].

Experimentation at the manufacturing stage is orders of magnitude more costly than experimentation at the R&D stage. As Kackar has pointed out [16],

> it is the designs of both the product and the manufacturing process that play crucial roles in determining the degree of performance variation and the manufacturing cost.

THE ROLE OF EXPERIMENTAL DESIGN

Data-based decisions, whether at the R&D level or at the manufacturing level, often require information that can be obtained most efficiently using designed experiments. This requires teamwork among researchers and statisticians. Researchers would agree that it is important for statisticians to understand the fundamentals of the production process. Statisticians would agree that it is important for researchers to understand the fundamentals of experimental design. For example, Snee has wryly observed [18],

> There are two important issues related to the implementation of the statistical approach to experimentation: getting scientists and engineers to (1) use statistical designs and (2) use the best possible design in a given situation. The majority of the benefit comes from using statistical designs in almost any form rather than worrying about finding the best possible design in each instance.

Collaboration between researchers and statisticians leads to well-planned experiments that can be interpreted clearly. As Hahn has pointed out [19],

> Results of a well planned experiment are often evident from simple graphical analyses. However, the world's best statistical analysis cannot rescue a poorly planned experimental program.

WHAT WE MUST DO

George Box has clearly stated the present position of American industry [20]:

> If we only follow, we must always be behind. We can lead by using statistics to tap the enormous reservoir of engineering and scientific skills available to us.... Statistics should be introduced ... as a means of catalyzing engineering and scientific reasoning by way of [experimental] design and data analysis. Such an approach ... will result in greater creativity and, if taught on a wide enough scale, could markedly improve quality and productivity and our overall competitive position.

REFERENCES

1. W. E. Deming, in W. A. Shewhart, *Statistical Method from the Viewpoint of Quality Control*, Department of Agriculture, Washington, DC, 1939, p. iv.
2. E. Fuchs, "Quality: Theory and practice," *AT&T Tech. J.*, **65**(2), 4–8 (1986).
3. A. B. Godfrey, "The history and evolution of quality in AT&T," *AT&T Tech. J.*, **65**(2), 9–20 (1986).
4. B. L. Joiner, "The key role of statisticians in the transformation of North American industry," *Am. Stat.*, **39**(3), 224–227 (1985). [See also "Comments" and "Reply," pp. 228–234.]
5. J. M. Juran, *Juran on Planning for Quality*, Macmillan, New York, 1988, p. 4–5.

6. *AT&T Statistical Quality Control Handbook*, 2nd ed., AT&T Technical Publications, Indianapolis, 1958. [Also known as the *Western Electric Statistical Quality Control Handbook*.]

7. W. E. Deming, *Quality, Productivity, and Competitive Position*, Massachusetts Institute of Technology, Cambridge, 1982, p. 22.

8. W. E. Deming, *Out of the Crisis*, Massachusetts Institute of Technology, Cambridge, 1986, p. 54.

9. R. W. Berger and T. H. Hart, *Statistical Process Control: A Guide for Implementation*, ASQC Quality Press, Milwaukee, 1986, p. 25.

10. R. W. Berger and T. H. Hart, *Statistical Process Control: A Guide for Implementation*, ASQC Quality Press, Milwaukee, 1986, p. 1.

11. E. L. Grant and R. S. Leavenworth, *Statistical Quality Control*, 5th ed., McGraw-Hill, New York, 1980.

12. G. E. P. Box, W. G. Hunter, and J. S. Hunter, *Statistics for Experimenters: An Introduction to Design, Data Analysis, and Model Building*, Wiley, New York, 1978, p. 1–15.

13. P. W. Holland, "Statistics and causal inference," *J. Am. Stat. Assoc.*, **81**, 945–960 (1986). [See also "Comments" and "Reply," pp. 961–970.]

14. S. N. Deming, "Optimization of experimental parameters in chemical analysis," in J. R. DeVoe, Ed., *Validation of the Measurement Process*, ACS Symposium Series, **63**, Washington, 1977, pp. 162–175.

15. G. Taguchi, *Introduction to Quality Engineering: Designing Quality into Products and Processes*, Kraus International Publications, White Plains, NY, 1986.

16. R. N. Kackar, "Off-line quality control, parameter design, and the Taguchi method," *J. Qual. Technol.*, **17**, 176–209 (1985).

17. M. Walton, *The Deming Management Method*, Dodd, Mead, New York, 1986, p. 131–157.

18. R. D. Snee, "Computer-aided design of experiments–some practical experiences," *J. Qual. Technol.*, **17**, 222–236 (1985).

19. G. J. Hahn, "Some things engineers should know about experimental design," *J. Qual. Technol.*, **9**, 13–20 (1977).

20. G. E. P. Box, "Signal-to-noise ratios, performance criteria, and transformations," *Technometrics*, **30**, 1–18 (1988).

Chapter **2**

Systems Theory and Response Surfaces

SINGLE-FACTOR SYSTEMS

A *system* is defined as a regularly interacting or interdependent group of items forming a unified whole. A system is described by its borders, by what crosses the borders, and by what goes on inside [1]. Figure 2.1 is a simple systems theory diagram [2] showing the three basic elements of systems theory: a factor (input), a response (output), and a transform (internal interrelationships between factors and responses).

A Simple Mathematical System

One example of a simple system is the mathematical relationship

$$y = 0.8x + 6.0 \tag{2.1}$$

shown schematically in Figure 2.2. The input to the system is the independent variable x (the factor). The output from the system is the dependent variable y (the response). The transform that relates the response to the factor is the mathematical relationship given in Equation 2.1. In the past, Equation 2.1 was an intangible example of a system, but with modern hand-held programmable calculators, the example has become more concrete: entering a value of 10 for x will produce a value of 14.0 for y.

Figure 2.1 Basic systems theory diagram showing a factor (input), a response (output), and the transform that relates changes in system responses to changes in system factors.

Figure 2.3 is a graph of the response y plotted as a function of the factor x for the system described by Figure 2.2 and Equation 2.1. The mathematical equation transforms a given value of the factor x into a value for the response y.

In this simple system, the transform is known with certainty. Thus, because we are given the transform of Equation 2.1, we can draw the graph shown in Figure 2.3. It is unnecessary to "experiment" with the system.

A Simple Chemical System

Figure 2.4 is another systems diagram showing a single factor (Fahrenheit temperature) and a single response (percentage yield). In this example, the transform is a set of chemical reactions that is responsible for converting raw materials (not shown) into usable product.

It should be pointed out that in this system, the transform between temperature and yield is probably *not* known with certainty. Although we might have a great deal of chemical theory available to us, and although we might have had much experience with this particular chemical system, it is nevertheless probably not possible to predict with high accuracy the percentage yield knowing only the Fahrenheit temperature. If we want to approximate the true transform with more and more certainty, it will be necessary to carry out experiments on the system.

It is a subtle but important point that in this chemical system the transform is not known with certainty. If our understanding of chemistry were perfect, we might be able to write a complete, exact model for the transform that would be the chemical equivalent of Equation 2.1. Unfortunately, our knowledge of chemistry is rarely that good and we must usually work with a model that only approximates the true transform.

This distinction between a model of the transform and the transform itself is important. The transform is the chemistry that goes on within the system, the

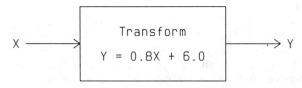

Figure 2.2 A systems theory diagram for the mathematical transform, $y = 0.8x + 6.0$.

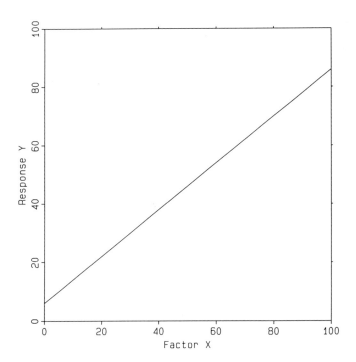

Figure 2.3 Graph of the mathematical system shown in Figure 2.2: $y = 0.8x + 6.0$. The intercept at $x = 0$ is 6.0; the slope (dy/dx) is 0.8.

actual chemical reactions and processes that convert the input materials into the output product. The model is an incomplete, approximate mathematical description of how we perceive the system to behave. In these modern days of computer chemistry and computer simulation, we have encountered students and professionals who sometimes fail to make this distinction between the transform and the model of the transform. They begin to believe that the model is reality, and forget that reality is the system itself. There *is* a fundamental difference [3].

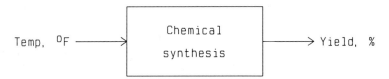

Figure 2.4 A systems theory diagram for a chemical transform. The single response (yield, %) is assumed to be a function of the single factor (temperature in °F).

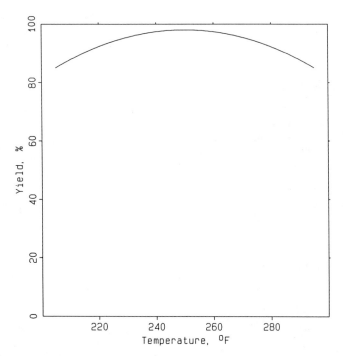

Figure 2.5 Graph of a relationship between percentage yield and temperature in °F. Note that high yield can be maintained over a relatively broad domain of temperature.

Ruggedness of Systems

Figure 2.5 is a graph of percentage yield plotted as a function of the Fahrenheit temperature for a system described by Figure 2.4. For this system, maximum yield (approximately 98%) can be achieved at a temperature of approximately 250°F. Perhaps more important is the fact that in the region of the optimum, the percentage yield does not change much as the temperature changes by a few degrees. As discussed in Chapter 1, such a system would be rugged: the response would be relatively insensitive to changes in the factor.

Suppose that the behavior of the system is not the behavior shown in Figure 2.5, but instead is represented by the graph in Figure 2.6. Maximum yield (approximately 98% again) can be achieved at a temperature of approximately 270°F in this example. But this example is not as rugged in the region of the optimum: changing the temperature by only a few degrees will cause spectacular decreases in the percentage yield.

Figures 2.5 and 2.6 suggest that optimization of a process is useful (often necessary), but most investigations are not complete without further experiments that show the ruggedness of the system with respect to the factors. Such informa-

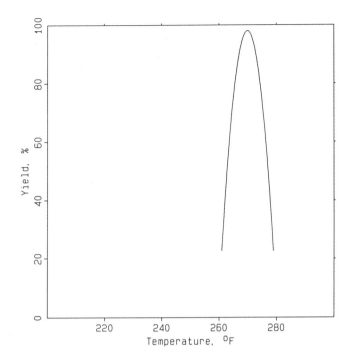

Figure 2.6 Graph of a relationship between percentage yield and temperature in °F. Note that high yield can be maintained over a relatively narrow domain of temperature.

tion lies behind the concept of "factor tolerances" [4–8], the type of information that should be passed from research to manufacturing to more rapidly bring products to market (see Chapters 1 and 9).

Noise

Figures 2.7 and 2.8 illustrate two types of uncertainty that are associated with response surfaces of real systems. Figure 2.7 shows *homoscedastic* noise, noise that is constant everywhere across the factor domain. In the upper graph, the smooth underlying response surface can be imagined with uniform noise superimposed. The lower graph plots the magnitude of this noise as a function of the factor *x*: by definition, the noise is constant.

Figure 2.8 shows *heteroscedastic* noise, noise that varies across the factor domain. In the upper graph, the smooth underlying response surface can be imagined beneath the varying amount of noise. The lower graph plots the magnitude of this noise as a function of the factor *x*: by definition, the noise is not constant.

Most real systems exhibit heteroscedastic noise when viewed over a large factor domain. One reason for this was considered in the discussion of Figures 1.5

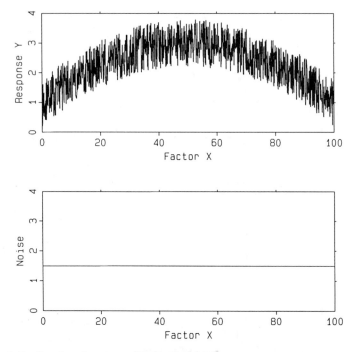

Figure 2.7 Graphs showing a homoscedastic response surface. Upper graph: smooth underlying response surface with uniform noise superimposed. Lower graph: magnitude of uniform noise as a function of the factor x.

and 1.6. For a given amount of variation in a factor x, the steepness of the response surface determines the amount of noise transformed into the response y. In the region of an optimum, the slope of the response surface is close to zero, so random fluctuations in the factor induce very little noise in the response. Far away from the optimum, however, the slope of the response surface is steeper (either positive or negative), and the same fluctuations in the factor give rise to larger amounts of noise in the response.

When viewed over a smaller factor domain, many real systems appear to exhibit homoscedastic noise. In Figure 2.8, for example, the noise between $x = 40$ and $x = 60$ can be considered to be constant for most practical purposes.

Most statistical procedures assume homoscedastic noise structures. If the noise is heteroscedastic, then special statistical techniques such as weighted least squares [9] can be used. Weighted least-squares methods place higher weight on information from regions of low noise and less weight on less certain information from regions of high noise. However, if experimental studies are confined to the region of the optimum, then in many cases the assumption of homoscedastic noise is appropriate, and special statistical techniques are not required. This is an additional (and somewhat esoteric) reason for reaching the region of the optimum before detailed experimental studies are carried out to characterize the response surface.

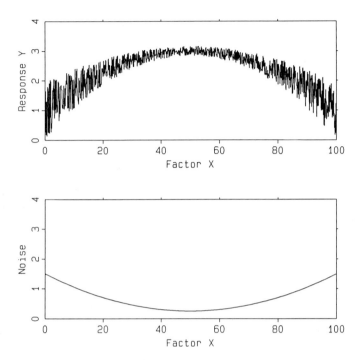

Figure 2.8 Graphs showing a heteroscedastic response surface. Upper graph: smooth underlying response surface with nonuniform noise superimposed. Lower graph: magnitude of nonuniform noise as a function of the factor x.

Measurement Systems

It is a common misconception that responses always come directly from the system of interest to the investigator. In most scientific and engineering fields, it is rare that the investigator can observe a system and obtain the response directly. Instead, measurement systems are used to observe the system of interest and transform information from one data domain to another [10]. Figure 2.9 shows the relationship between a primary system under investigation and an associated measurement system.

The result we "see," the measured response, has come through the measurement system. Any bias or noise contributed by the measurement system will be added to the information coming from the system of interest to the investigator. Ideally, a measurement process will be free of bias and will contribute zero noise [11]. In practice, bias is often less of a problem than long- and short-term noise. A general rule is that the measurement system should have an order of magnitude less noise than the noise in the system of interest to the investigator. Experimental designs exist for determining the components and sources of noise [12].

Perhaps not coincidentally, analytical chemists (who make measurements on chemical systems) have been among the most frequent users of optimization and experimental design techniques [13] to minimize bias and reduce noise.

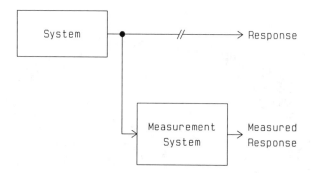

Figure 2.9 Diagram showing how systems are observed. The response does not come directly from the system of interest, but instead is almost always filtered through a measurement system.

MULTIFACTOR SYSTEMS

Figure 2.10 is a systems theory diagram that is more representative of most real systems. Systems generally have multiple factors (x_1, x_2, x_3, ...), multiple responses (y_1, y_2, y_3, ...), and considerable subsystem structure. In this chapter we will consider systems that have only a single response, y_1. Later, in Chapter 8, multiresponse systems will be considered.

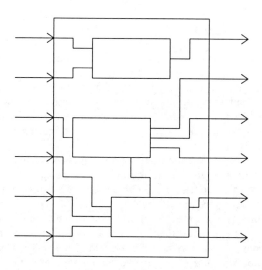

Figure 2.10 A more realistic systems theory view of most processes showing multiple factors, multiple responses, and considerable subsystem structure.

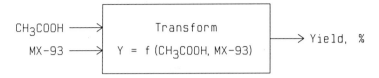

Figure 2.11 A systems theory view of a two-factor chemical process. The response (yield, %) is a function of the two factors, acetic acid (CH₃COOH) and a proprietary ingredient, MX-93.

A Simple Two-Factor Chemical System

Figure 2.11 is a systems theory view of a chemical process for which there are two factors (x_1 = CH₃COOH, acetic acid; x_2 = MX-93, a proprietary ingredient) and a single response (y_1 = percentage yield of desired product). The transform between the response and factors is probably based on chemistry, but the exact form of this relationship is not usually known with certainty. Thus, the transform is represented as

$$y_1 = f(\text{CH}_3\text{COOH}, \text{MX-93}) \qquad (2.2)$$

where f is some unknown function. Because the relationship between the response (percentage yield) and the factors (acetic acid and MX-93) is not known with certainty, the approximate form of f must be discovered by experiment.

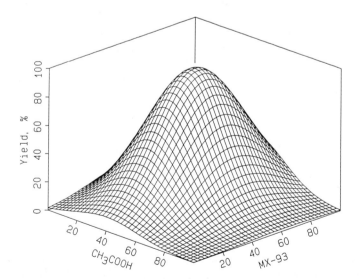

Figure 2.12 Graph of percentage yield as a function of the two system factors, CH₃COOH and MX-93, for the chemical process shown in Figure 2.11.

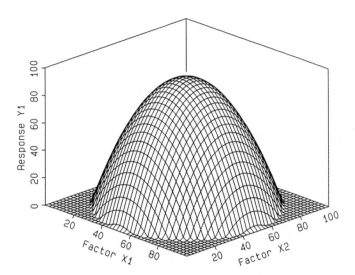

Figure 2.13 Graph of response y_1 as a function of factors x_1 and x_2: $y = -107.00 + 4.00x_1 + 4.00x_2 - 0.0400x_1^2 - 0.0400x_2^2 + 0.00x_1x_2$. The response surface is centered at $x_1 = 50.0$, $x_2 = 50.0$ and is symmetric about this point.

Figure 2.12 is a graph of the system response y plotted as a function of the two system factors, x_1 and x_2. The chemical transformation converts combinations of CH_3COOH and MX-93 into yield of desired product. There is some combination of CH_3COOH and MX-93 that will give maximal yield; all other combinations of these two factors will produce lower response.

Again, it must be remembered that views such as that given in Figure 2.12 are not generally available to the researcher at the beginning of an investigation. This information must be discovered by experiment. A goal of many projects is to find the combination of factors that gives the optimum response.

Some Examples of Two-Factor Response Surfaces

The following three response surfaces are based on full two-factor second-order polynomial models of the form

$$y_1 = \beta_0 + \beta_1 x_1 + \beta_2 x_2 + \beta_{11} x_1^2 + \beta_{22} x_2^2 + \beta_{12} x_1 x_2 + r_1 \tag{2.3}$$

where y_1 is the response, x_1 and x_2 are the factors, and the βs are the parameters of the model. This type of model is often used to approximate real response surfaces over a limited domain of factor space (see Chapter 9). Responses less than zero are set equal to zero in these figures.

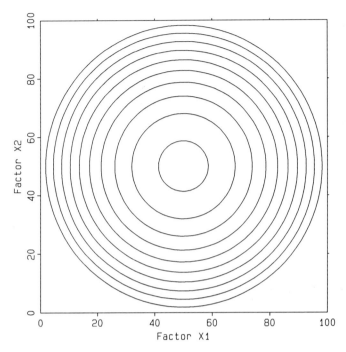

Figure 2.14 Contours of constant response for the response surface shown in Figure 2.13. Contours are shown for $y_1 = 90.0, 80.0, 70.0, 60.0, 50.0, 40.0, 30.0, 20.0, 10.0$, and 0.00. The circular contours are symmetric about the point $x_1 = 50.0$, $x_2 = 50.0$.

Figure 2.13 is a pseudo-three-dimensional plot of a two-factor response surface. It shows how the response, y_1, varies as a function of the two factors, x_1 and x_2. There is one combination of x_1 and x_2 that will produce maximal response from the system: $x_1 = 50.0$, $x_2 = 50.0$, $y_1 = 93.0$, in this example. All other combinations of x_1 and x_2 will produce lower responses.

Imagine that we can secure a huge knife to the response axis in Figure 2.13. The knife is fastened in such a way that it will pivot around the y axis and cut through the response surface parallel to the x_1–x_2 plane. If the knife is adjusted so the cutting edge is positioned at a response of 90 and is swung around the y axis, it will slice through the response surface. Suppose we climb onto the surface, paint the knife cut with black paint all the way around, and climb down again. We repeat the procedure with the knife positioned at responses of 80, 70, 60, 50, 40, 30, 20, 10, and 0. When we finish, we'll have a mountain with 10 black stripes running around it at different heights.

Imagine now that we rent a helicopter and fly directly above the mountain. When we look down, we will have lost all perception of depth, but we will see the knife cuts that we painted black. They will form lines against the otherwise featureless surface. Our view will be that shown in Figure 2.14. This *contour plot*

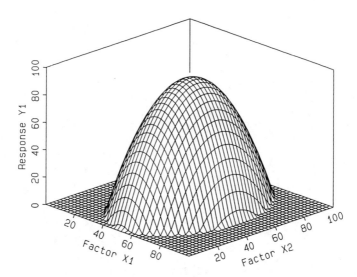

Figure 2.15 Graph of response y_1 as a function of factors x_1 and x_2: $y = -207.00 + 8.00x_1 + 4.00x_2 - 0.0800x_1^2 - 0.0400x_2^2 + 0.00x_1x_2$. The response surface is centered at $x_1 = 50.0$, $x_2 = 50.0$ and is symmetric about a vertical plane at $x_1 = 50.0$.

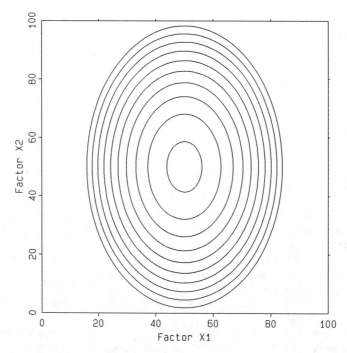

Figure 2.16 Contours of constant response for the response surface shown in Figure 2.13. Contours are shown for $y_1 = 90.0, 80.0, 70.0, 60.0, 50.0, 40.0, 30.0, 20.0, 10.0$, and 0.00. The elliptical contours are symmetric about the line $x_1 = 50.0$.

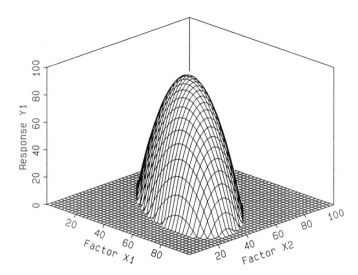

Figure 2.17 Graph of response y_1 as a function of factors x_1 and x_2: $y = -666.81 + 12.2x_1 + 18.2x_2 - 0.0700x_2^2 - 0.130x_1^2 - 0.104x_1x_2$. The response surface is centered at $x_1 = 50.0$, $x_2 = 50.0$ and is symmetric about a vertical plane passing through that point and making an angle of $120°$ with the x_1 factor axis.

shows fixed values of response, y_1, as a function of the two factors, x_1 and x_2. We can deduce the shape of the response surface from the *contours of constant response*. The circular contours of constant response suggest that the response surface is symmetrical about the point $x_1 = 50.0$, $x_2 = 50.0$, and that the response surface decreases more and more rapidly as we move farther and farther away from this central point.

Circularly symmetric response surfaces such as that represented by Figures 2.13 and 2.14 are easy to optimize. Unfortunately, they occur very infrequently.

Figure 2.15 is another response surface showing how the response y_1 varies as a function of the two factors, x_1 and x_2. Again, there is one combination of x_1 and x_2 that will produce maximal response from the system ($x_1 = 50.0$, $x_2 = 50.0$, $y_1 = 93.0$). All other combinations of x_1 and x_2 will produce lower responses. However, note that the response surface falls off more rapidly in the x_1 direction than it does in the x_2 direction. This is confirmed in Figure 2.16, which shows contours of constant response. The elliptical nature of these contours (each major axis is parallel with the x_2 axis, and each minor axis is parallel with the x_1 axis) shows that small changes in x_1 produce greater decreases in response than the same small changes in x_2.

Although the effect of factor x_1 is not the same as the effect of factor x_2, their effects are independent. Figures 2.15 and 2.16 also represent response surfaces that are easy to optimize, but occur infrequently.

Figure 2.17 is a third response surface showing how the response y_1 varies as a function of the two factors, x_1 and x_2. There is one combination of x_1 and x_2 that will produce maximal response from the system ($x_1 = 50.0$, $x_2 = 50.0$, $y_1 = 93.0$). All other combinations of x_1 and x_2 will produce lower responses.

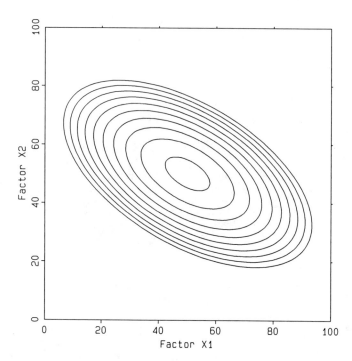

Figure 2.18 Contours of constant response for the response surface shown in Figure 2.13. Contours are shown for y_1 = 90.0, 80.0, 70.0, 60.0, 50.0, 40.0, 30.0, 20.0, 10.0, and 0.00. The elliptical contours are symmetric about a line passing through x_1 = 50.0, x_2 = 50.0 and making an angle of 120° with the x_1 factor axis.

Note that the response falls off more rapidly in one direction than it does in another, perpendicular direction, but these directions are no longer oriented parallel with the factor axes. If the x_1 factor axis is oriented to run west to east, the response decreases from the optimum most rapidly in the southwest–northeast direction, and more slowly in the northwest–southeast direction. The direction of steepest decrease is not parallel with either of the factor axes, but is instead oblique with reference to them. This is confirmed in Figure 2.18 which shows contours of constant response.

As we will see, Figures 2.17 and 2.18 represent response surfaces that are more difficult to optimize. In most scientific and engineering fields, these oblique ridge systems are usually the rule rather than the exception [14–16].

Figure 2.19 shows *ellipsoidal* (football-like) contours of constant response for a three-factor system (note that all three orthogonal axes in this figure are factor axes; the response is not shown). Any combination of the three factors that lies on the outermost ellipsoidal surface might give a response of 70. Any combination of the three factors that lies on the middle ellipsoidal surface might give a response of 80. Any combination of x_1, x_2, and x_3 that lies between these two ellipsoidal surfaces would give a response between 70 and 80. The optimum combination of x_1, x_2,

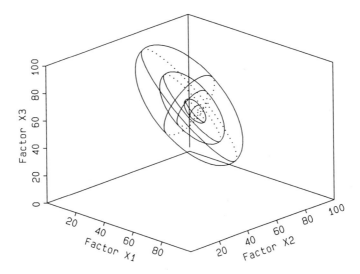

Figure 2.19 Ellipsoidal contours of constant response as a function of three factors, x_1, x_2, and x_3. The optimum combination of x_1, x_2, and x_3 lies at the center of the innermost ellipsoid. All other combinations of x_1, x_2, and x_3 give less than optimum responses.

and x_3 lies at the center of the innermost ellipsoid. All other combinations of x_1, x_2, and x_3 give less than optimum responses. Additional three-factor contours of constant response are given in Box, Hunter, and Hunter [17].

OPTIMIZATION STRATEGIES

Single-Factor Optimization

There are many approaches to single-factor optimization [18]. Figure 2.20 shows one efficient and effective approach [19].

It was desired to find the pH for which the rate of an enzyme-catalyzed reaction would be maximum. The pH was varied by adding a constant total amount, but different relative amounts, of pH 8.72 and 10.72, 0.837 M 2-amino-2-methyl-1-propanol buffers. The algorithm for determining the pH maximum required five initial experiments in which the ratio of buffer volumes was varied in four increments to cover the range of interest of the independent variable (pH 8.72 to 10.72). These data were examined to find the maximum rate in the set. Two new experiments were designed and carried out at intervals equal to one-half the original interval on either side of the volume ratio corresponding to the maximum rate. The procedure was repeated until the resolution of the independent variable (buffer ratio) was 1/32 of the range. This required 11 experiments to attain this resolution at the pH maximum. The results are shown in Figure 2.20. The numbers beside the data points show the experiment order.

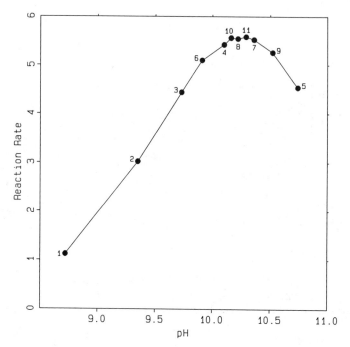

Figure 2.20 A single-factor search to find optimal pH. Numbers indicate experiment order. See text for details.

This single-factor optimization strategy has several advantages:

1. Only a few experiments (11) were needed to get into the region of the optimum.
2. The density of experiments is greatest in the region of the optimum. Most of the experiments were carried out in the region of ultimate interest.
3. Those regions that are not of much interest have not been explored unnecessarily. Sparse information has been obtained in those regions, just enough to know that additional experiments there are probably not justified.
4. The ruggedness of the system is readily apparent from the well-defined curvature in the region of the optimum.

Efficient and effective optimization strategies such as that shown in Figure 2.20 are not found very frequently in research and development projects. Instead, projects get bogged down because scientists and engineers try to gain a complete understanding of the system in narrow regions that are not of ultimate interest. Far too often we have seen applied research groups carrying out the equivalent of detailed investigations of the factor domain between pH 9.00 and 9.50 when their real objective is to find the pH at which the reaction rate will be optimal. This is wasteful, inefficient, and misdirected [20].

If you can use rapid optimization techniques such as that shown in Figure 2.20, while your competition still tries to gain a complete understanding of the

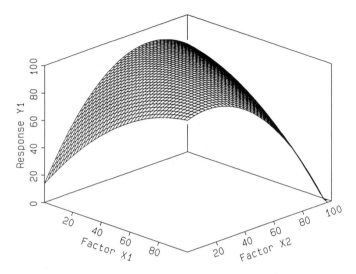

Figure 2.21   A two-factor response surface.

system in narrow regions that are not of ultimate interest, you will beat the competition faster and with less effort.

Stated differently, if your competition can use rapid optimization techniques such as that shown in Figure 2.20, while you still try to gain a complete understanding of the system in narrow regions that are not of ultimate interest, then your competition will beat you faster and with less effort.

Multifactor Optimization

Figure 2.21 shows a two-factor response surface – response y_1 as a function of factors x_1 and x_2. There are three common approaches that are used by researchers to optimize multifactor systems such as this.

The Theoretical or Consultant's Approach.   Scientists and engineers have developed theoretical models that frequently do a reasonable job of predicting system behavior. Thus, if asked to predict conditions that will give maximal yield in a chemical reaction, it is sometimes possible to use theoretical models to suggest conditions that will give a good response. And the models are often right.

The dot in Figure 2.22 shows the location of a single experiment carried out at the theoretically optimum combination of x_1 and x_2. Evidently the theory is approximately correct, but the real optimum is located elsewhere. Some researchers might be pleased with the good yield at the factor combination shown in Figure 2.22. If they do not do any more experiments, they might conclude that they have indeed found the optimum.

Keep in mind that the investigator usually does not know what the response surface looks like when the investigation is begun – the shape has to be discovered through a series of experiments. One experiment, by itself, gives no information

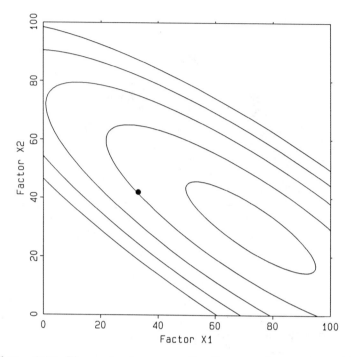

Figure 2.22   The academic or consultant's approach to optimization.

about the shape of the response surface. Shape is a differential quality, dy/dx. To define shape, there must be a dx. This in turn requires at least two different values of x. If you want to find out how a factor influences a response, you must vary that factor.

Nonetheless, if no prior experimental information exists, the theoretical approach is a superb strategy to get a project started.

The Shotgun Approach.   Many of us are familiar with the shotgun approach to multifactor experimentation (see Figure 2.23). Other descriptive names for this are the "stochastic strategy" and the "probabilistic strategy" (these terms are preferred for monthly reports).

One advantage of a totally random shotgun approach is that if enough experiments are carried out, they will cover the factor space well. This can be important if information is needed over the entire factor domain. An example is in basic research where understanding the underlying mechanism is desired. Another example is in applied research where you must have some confidence that your competition has not found an operating region that is better than the one you are using.

However, the totally random shotgun approach is inefficient because it will eventually reexplore regions that have already been explored. For example, the chance cluster of data points at the bottom left of Figure 2.23 is unfortunate. Only one experiment is enough to find out that the response is undesirable and further investigation is not needed.

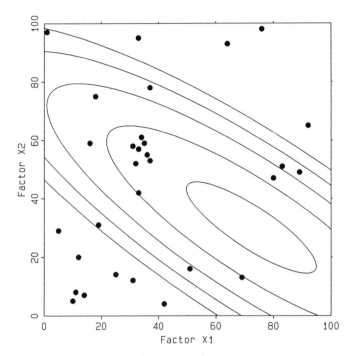

Figure 2.23 The "shotgun" (or "stochastic" or "probabilistic") approach to optimization.

In the "directed shotgun" approach, future experiments are carried out in regions that have been found to give promising results. But this approach is also inefficient. An example is shown by the cluster of data points in Figure 2.23 near the coordinates $x_1 = 35$, $x_2 = 55$. Only a few of these experiments are necessary to demonstrate that this is a good region. Any additional experiments will consume time, money, and resources but will not provide any new information. Even worse, however, is that the attractiveness of this region has kept us from exploring other regions and finding the true optimum. This is sometimes referred to as the "lamp post effect."

There is an old story about a researcher who was walking home late one night. As he approached a street corner, he saw a man crawling around on his hands and knees under the lamp post. The researcher asked the man what he was doing, and in an obviously inebriated voice, the stranger replied, "Jush lookin' fer m' keys." The researcher felt sorry for the man, and got down on his hands and knees to help look for the keys. After about ten minutes the researcher asked the man if he was sure he had lost his keys here. "Naw," said the stranger, "I los' 'em over there in th' bushes." "Well for heaven's sake, why are you looking for them here?" asked the researcher. "'Cause th' light's better here," was the reply.

Unfortunately, much research is carried out "under the lamp post" where the light is better. We tend to do our experiments in regions that have already had light

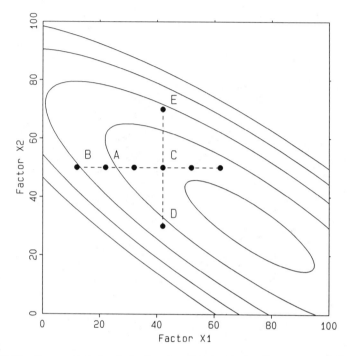

Figure 2.24 Sequential single-factor-at-a-time optimization on a multifactor surface. See text for details.

shed on them by us or by others. We seldom try factor combinations very far removed from where we have already been operating.

But probably the biggest difficulty with "shotgun" experiments is that the results can be confusing and hard to interpret. Many of us are used to asking (and being asked) naive questions such as, "What is the effect of the factor x_1 on the response? Does increasing the value of x_1 cause the response to go up, or does increasing the value of x_1 cause the response to go down?" The questioner expects a simple, naive, unqualified, either–or answer.

The problem here is that both answers are correct. It is clear from Figure 2.23 that at low values of x_2, increasing the value of x_1 causes the response to go up. But at high values of x_2, increasing the value of x_1 causes the response to go down. Thus, the answer must be qualified—we must state the conditions of x_2 for which our chosen answer applies.

When the effect of one factor depends on the level of another factor, an *interaction* exists between the two factors. Most real systems exhibit factor interaction; interaction is the rule rather than the exception in the real world.

The Single-Factor-at-a-Time Approach. Figure 2.24 shows a "multifactor strategy" that is often employed by researchers. It is called the "single-factor-at-a-time approach."

Sometime during your scientific education you were probably given the following advice about what constitutes Good Science: "If you want to find out how a factor affects the response, you must hold all other factors constant and vary that one factor only." It sounds good. But you were not told that the answer you get is conditional on the values of all the other factors. As we saw in Figure 2.23, if we change the value of one of these other factors, we will probably get a different answer because of factor interaction.

Many researchers are not aware of factor interaction. They believe they can optimize a multifactor system by carrying out a series of single-factor-at-a-time studies. Their results are necessarily limited.

Consider the sequence of experiments shown in Figure 2.24. To do traditional "good science," the researcher has decided to hold factor x_2 constant and vary only factor x_1. An initial experiment is carried out at point A. A second experiment is carried out at the same value of x_2 but at a lower level of x_1, at point B. These two points establish a gradient that increases along factor x_1. Additional experiments are carried out at increasing levels of the factor x_1 (always holding x_2 constant) until the response begins to decrease. From the results of this set of experiments varying x_1, point C is chosen as the optimum. The researcher has "optimized factor x_1."

Continuing to do "good science," the researcher now holds factor x_1 constant at its optimum level and varies factor x_2. Experiment D suggests that the gradient increases in the direction of increasing values of x_2, but experiment E contradicts this. Further experiments between D and E would reveal that point C is the optimum response. The researcher has now "optimized factor x_2."

Close inspection reveals that the optimum value of x_2 is the value of x_2 chosen initially by the experimenter. Some researchers notice this, and say, "See! I knew ahead of time what the optimum value of factor x_2 should be. What a brilliant theoretician I am." Perhaps. But what has happened in Figure 2.24 is that the single-factor-at-a-time optimization strategy has become stranded on the oblique ridge. The only way to move along this ridge toward the optimum is to change all factors simultaneously. But the single-factor-at-a-time strategy does not change all factors simultaneously. A true multifactor optimization strategy is needed.

EVOLUTIONARY OPERATION (EVOP)

Engineers have certain fantasies about the ideal process: it would be an unchanging process, a process that has "lined out," that always produces the desired output, day after day after day. In an attempt to achieve this ideal process, engineers often try to keep the process operating at some fixed set-point. The set-point is almost never changed. Many processes are run "with the knobs taped down." There are no "adjustable variables." The philosophy seems to be, "If it ain't broke, don't fix it."

Unfortunately, many processes are operating delicately on the sides of response surface hills [21]. Small changes in the system factors cause large changes in the system response (see Figure 1.5). Most processes need improvement, not only to improve the responses but also to make the processes more rugged. However, the lamp post effect, a tape-the-knobs-down attitude, an if-it-ain't-broke-don't-fix-it philosophy, all hamper the needed improvements.

George Box viewed processes differently [16, 22]. He viewed processes as experimental systems capable of being changed. The problem was that engineers always wanted to do the same experiment over and over again. But they could never

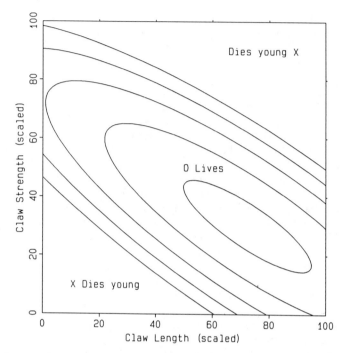

Figure 2.25 Lobster survival contours as a function of claw length and claw strength.

learn anything about the effects of the experimental factors that way. (After all, if you want to find out how a factor influences a response, you must vary that factor.)

Box suggested changing the experimental conditions very slightly. By making very small changes in the factors, there would be no large changes in response (the product could still be sold), but the small changes in response that did occur could be monitored. The results of these small changes in the factors would reveal information about how the system should be changed to improve the response.

Box [16, 22] introduced this concept with an analogy to the survival of adolescent lobsters (Figure 2.25). Two factors that affect the lobster's survival rate are claw length and claw strength. If the lobster's claw is too short, the lobster cannot grab its opponent very well, and might die young before it can reproduce. If the claw is too long, the lobster would be too cumbersome and slow, and might also die young before it can reproduce. There is probably some optimum claw length for an adolescent lobster. The same is probably true for claw strength. If the lobster's claw is too weak, the lobster cannot defeat its opponent, and might die young. If the lobster's claw is too strong, it might defeat its current opponent but might also damage its own claw in the process and fight poorly in its next battle, so poorly that it might die young.

Figure 2.25 shows survival contours for adolescent lobsters as a function of claw length and claw strength. If a lobster has a highly effective combination of claw length and claw strength, that lobster will probably live long enough to

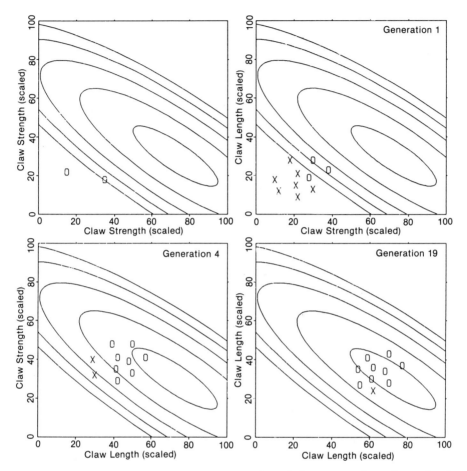

Figure 2.26 Evolution of lobster claw length and claw strength to the optimal combination. See text for details.

reproduce. The characteristics of such a lobster are marked by the symbol O in Figure 2.25. If a lobster has a highly ineffective combination of claw length and claw strength, the lobster will probably die young and not be able to reproduce. The characteristics of such lobsters are marked by the symbol X in Figure 2.25.

Now, consider a mother and father lobster who have undesirable combinations of claw length and claw strength, but have nonetheless managed to survive long enough to reproduce. The characteristics of these two lobsters are shown in the upper left panel of Figure 2.26. The characteristics of their children are shown as symbols in the upper right panel of Figure 2.26. Note that the average claw length and average claw strength of all of the children are about the same as the average claw length and average claw strength of the parents. Note also that there is considerable variation in these characteristics among the children.

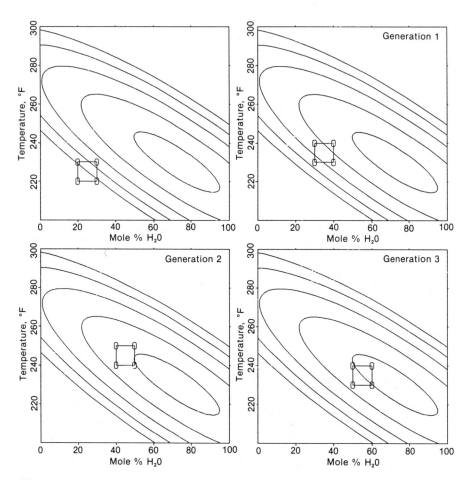

Figure 2.27 Box-type *ev*olutionary *op*eration (EVOP) using a two-level two-factor full factorial experimental design. See text for details.

Those children whose characteristics are less well suited for their environment run a greater risk of dying young than do their better suited brothers and sisters. The characteristics of the progeny who die before they can reproduce are indicated by Xs in Figure 2.26; the characteristics of those who live long enough to reproduce are indicated by Os. This is straightforward "survival of the fittest" [23]. Note that the average claw length and average claw strength of the *surviving* lobsters have moved toward a more desirable combination. This natural selection has caused the surviving lobster population to evolve toward the optimum.

Presumably, the progeny of the next generation will have characteristics of claw length and claw strength that are, on the average, the same as their parents. But with further natural selection of the better suited individuals, the average characteristics of the surviving population will move even more closely toward the optimum. Figure 2.26 shows the fourth and nineteenth generations of lobsters. In this example, the lobsters have evolved their claw length and claw strength to achieve optimal survival.

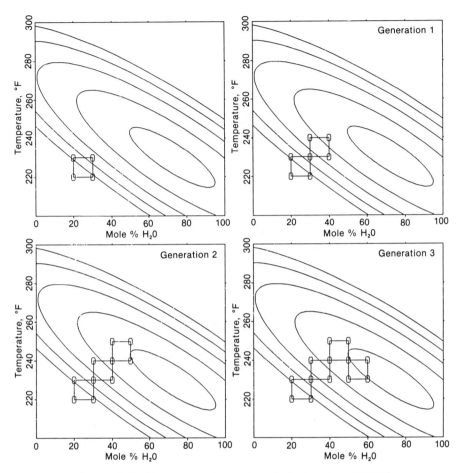

Figure 2.28 Summary of experiments for box-type EVOP.

Box coined the term "evolutionary operation," or EVOP for short, to describe a process whereby systems other than lobsters can evolve to their optimum contours of constant response. The major ideas behind EVOP are as follows:

1. Lobster evolution concepts are applicable to response surfaces other than survival surfaces.
2. Claw length and claw strength can be replaced with factors associated with chemical (or other) processes, factors such as temperature and mol% H_2O.
3. Replace random natural variation with orderly, statistically based variation. Classical factorial designs were chosen, primarily because they allow estimates of interactions among factors.
4. From the results of the factorial experiments, determine the direction to move.
5. "Kill off" all experimental conditions except the ones that produced the most desirable result.
6. Create a new generation of experiments in the desirable direction.

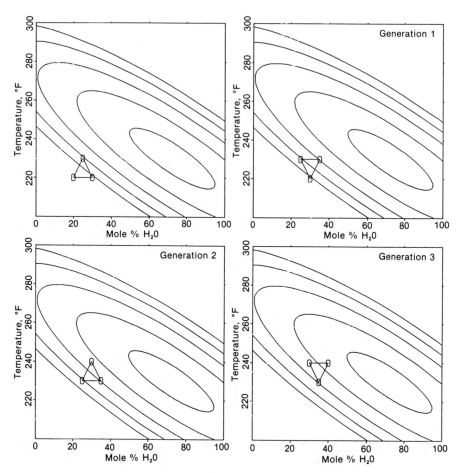

Figure 2.29 Simplex EVOP using a two-factor simplex experimental design. See text for details.

These ideas are shown in Figure 2.27. The initial two-factor (x_1 and x_2), two-level (low and high in each factor: 20 and 30 mol% H_2O; 220 and 230°F) full factorial design is shown in the upper left panel of Figure 2.27. This is often designated a 2^2 factorial design, and has $2^2 = 4$ distinctly different factor combinations. In a real industrial process, the changes in the factors might be much less than the changes shown in Figure 2.27. One of the factor combinations in the initial factorial design would probably be the current set-point of the operating process.

Information obtained from the initial factorial design shown in the upper left panel of Figure 2.27 indicates that increased response might be obtained by increasing the mol% H_2O and by increasing the temperature. Because the factorial design finds itself located on the more or less flat side of the response surface, there is little interaction among factors here. By "killing off" all but the upper right experiment (30 mol% H_2O, 230°F), and constructing a new factorial design anchored to this remaining point with increases in both mol% H_2O and tempera-

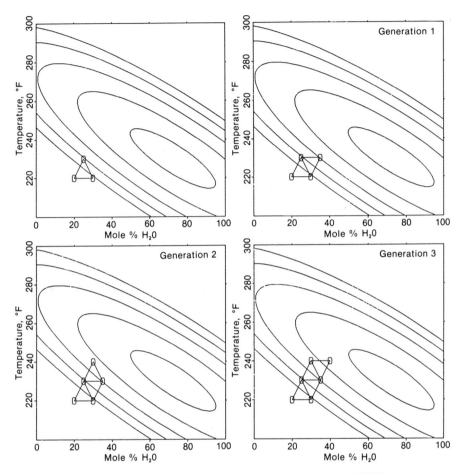

Figure 2.30 Summary of experiments for simplex EVOP.

ture, the second-generation factorial design shown in the upper right panel of Figure 2.27 is obtained.

Further application of this EVOP strategy produces the experimental designs shown in the remaining two panels of Figure 2.27. The cumulative results are summarized in Figure 2.28. Hunter and Kittrell have reviewed the use of factorial EVOP [24].

SIMPLEX EVOP

W. Spendley, G. R. Hext, and F. R. Himsworth published a paper in 1962 [25] in which they set out, among other things, to make EVOP automatic:

> The work now to be described sprang originally from consideration of how Evolutionary Operation might be made automatic—a possibility to which Box drew attention even in 1955. More specifically, two problems were

posed, viz. (i) could an Evolutionary Operation procedure be devised which would more rapidly approach and attain optimum conditions, and (ii) could the calculations and decisions be so formalised and simplified that they could be executed automatically by a digital computer or–analogously–by a plant manager with little time to spare from routine duties?...

The basic design of the technique is the regular simplex in k dimensions, k being the number of factors or variables under study.

A simplex is a geometric figure having a number of vertexes (corners) equal to one more than the number of dimensions of the factor space. Thus, in two-factor dimensions, the simplex has $k + 1 = 3$ vertexes and is a triangle. The simplex represents a tangential planar approximation to the response surface in the region of the design. Spendley et al. [25] proposed using this simplex pattern instead of the factorial pattern of Box [16, 22] in an evolutionary scheme. Figures 2.29 and 2.30 are the simplex EVOP analogies of the factorial EVOP schemes shown in Figures 2.27 and 2.28. In Figure 2.30 note that a new simplex can be formed by eliminating the vertex having the worst response in the set of three, projecting its coordinates through the average coordinates of the remaining two vertexes an equal distance beyond, and carrying out only one new experiment at this reflection vertex.

Simplex EVOP has two major advantages over classical factorial EVOP. First, the number of experiments in the initial simplex design is only $k + 1$ instead of 2^k in the initial factorial design. This is to say that the number of experiments in the initial simplex design increases arithmetically, whereas the number of experiments in the initial factorial design increases geometrically. Comparisons of the number of experiments for increasing numbers of dimensions are as follows:

Dimensions	Simplex experiments	Factorial experiments
1	2	2
2	3	4
3	4	8
4	5	16
5	6	32

The second major advantage the simplex has over classical factorial EVOP is that the simplex requires only one new experiment to move into an adjacent region of factor space whereas the factorial design requires (at best) one-half the number of experiments in the factorial design.

The sequential simplex is discussed in more detail in the following chapters.

REFERENCES

1. S. N. Deming and S. L. Morgan, *Experimental Design: A Chemometric Approach*, Elsevier, Amsterdam, 1987.
2. L. von Bertalanffy, *General System Theory: Foundations, Development, Applications*, George Braziller, New York, 1968.

3. I. Asimov, *The Relativity of Wrong*, Oxford University Press, Oxford, 1989, pp. 213–225.

4. A. L. Wilson, "The performance-characteristics of analytical methods – I," *Talanta*, **17**, 21–29 (1970).

5. A. L. Wilson, "The performance-characteristics of analytical methods – II," *Talanta*, **17**, 31–44 (1970).

6. A. L. Wilson, "The performance characteristics of analytical methods – III," *Talanta*, **20**, 725–732 (1973).

7. P. J. Ross, *Taguchi Techniques for Quality Engineering: Loss Function, Orthogonal Experiments, Parameter and Tolerance Design*, McGraw-Hill, New York, 1988.

8. G. E. P. Box, "Signal-to-noise ratios, performance criteria, and transformations," *Technometrics*, **30**, 1–18 (1988).

9. W. E. Deming, *Statistical Adjustment of Data*, Dover, New York, 1943.

10. C. G. Enke, "Data domains – an analysis of digital and analog instrumentation systems and components," *Anal. Chem.*, **43**, 69–77 (1971).

11. J. K. Taylor, *Quality Assurance of Chemical Measurements*, Lewis Publishers, Chelsea, MI, 1987.

12. J. R. Smith and J. M. Beverly, "The use and analysis of staggered nested factorial designs," *J. Quality Tech.*, **13**, 166–173 (1981).

13. A. B. Hoadley and J. R. Kettenring, "Communications between statisticians and engineers/physical scientists," *Technometrics*, **32**, 243–247 (1990). [See also "Commentary," pp. 249–274.]

14. G. E. P. Box and K. B. Wilson, "On the experimental attainment of optimum conditions," *J. R. Stat. Soc., Ser. B*, **13**, 1–45 (1951).

15. G. E. P. Box, "The exploration and exploitation of response surfaces: Some general considerations and examples," *Biometrics*, **10**, 16–60 (1954).

16. G. E. P. Box and N. R. Draper, *Evolutionary Operation: A Method for Increasing Industrial Productivity*, Wiley, New York, 1969.

17. G. E. P. Box, W. G. Hunter, and J. S. Hunter, *Statistics for Experimenters: An Introduction to Design, Data Analysis, and Model Building*, Wiley, New York, 1978.

18. G. S. G. Beveridge and R. S. Schechter, *Optimization: Theory and Practice*, McGraw-Hill, New York, 1970.

19. S. N. Deming and H. L. Pardue, "An automated instrumental system for fundamental characterization of chemical reactions," *Anal. Chem.*, **43**, 192–200 (1971).

20. R. M. Driver, "Statistical methods and the chemist," *Chem. Br.*, **6**, 154–158 (1970).

21. S. N. Deming, "Optimization," *J. Res. Natl. Bur. Stand.*, **90**(6), 479–483 (1985).

22. G. E. P. Box, "Evolutionary operation: A method for increasing industrial productivity," *Applied Statistics*, **6**, 81–101 (1957). [Paper presented to the International Conference on Statistical Quality Control, Paris, July 1955.]

23. C. Darwin, *The Origin of Species by Means of Natural Selection: or the Preservation of Favored Races in the Struggle for Life*, Appleton, New York, 1899.

24. W. G. Hunter and J. R. Kittrell, "Evolutionary operation: A review," *Technometrics*, **8**, 389–397 (1966).

25. W. Spendley, G. R. Hext, and F. R. Himsworth, "Sequential application of simplex designs in optimisation and evolutionary operation," *Technometrics*, **4**, 441–461 (1962).

Chapter **3**

The Basic Simplex Algorithm

DEFINITION OF A SIMPLEX

A simplex is a geometric figure that has a number of vertexes (corners) equal to one more than the number of dimensions in the factor space.

If k is the number of dimensions in the factor space, then a simplex is defined by $k+1$ points in that factor space.

As shown in Figure 3.1, a two-factor simplex is defined by three points. In two dimensions, a simplex is a triangle.

When using a simplex for the optimization of experimental systems, each vertex corresponds to a set of experimental conditions. The solid lines drawn between vertexes are used to visualize the simplex. They have no other function. It is the vertexes only that determine experimental conditions. In Figure 3.1, for example, the first experiment would be carried out at a pH of 2.0 and a temperature of 25°C, the second experiment would be carried out at a pH of 4.5 and a temperature of 30°C, and the third experiment would be carried out at a pH of 3.0 and a temperature of 35°C. In tabular form:

Experiment	pH	Temperature	Result
1	2.0	25	—
2	4.5	30	—
3	3.0	35	—

The results of the experiments in this simplex would be tabulated and used to determine the next set of experimental conditions.

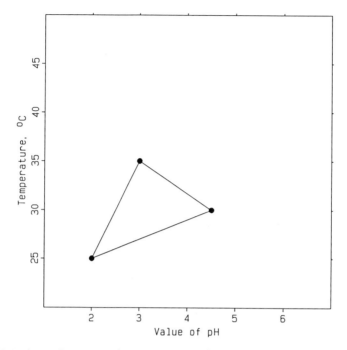

Figure 3.1 A two-factor simplex is represented as a triangle in a two-dimensional plane. Each vertex (corner) corresponds to a set of experimental conditions.

SIMPLEXES IN VARIOUS DIMENSIONS

As shown in Figure 3.2, a simplex can be defined for any number of factors, even for no factors.

Zero Dimensions

Because a simplex is a geometric figure that has a number of vertexes equal to one more than the number of dimensions in the factor space, in zero dimensions the simplex must have one vertex. The geometric figure that has only one vertex is the single point (Figure 3.2A). However, a zero-dimensional simplex cannot be made to move because it has no dimensions in which to move. It has no degrees of freedom. A zero-dimensional simplex is an admittedly trivial case, and is not very useful.

One Dimension

By definition, a simplex in one dimension must have two vertexes. The geometric figure that has only two vertexes is the line segment (Figure 3.2B). A line segment can be made to move, but it can move in only one direction, the direction of the

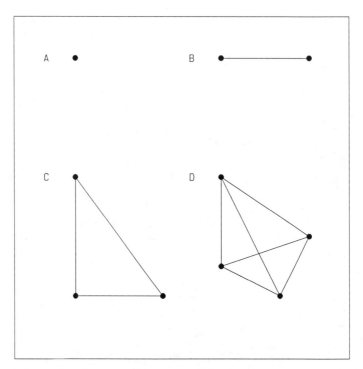

Figure 3.2 Simplexes in (A) zero-dimensional, (B) one-dimensional, (C) two-dimensional, and (D) three-dimensional factor spaces. Higher dimensional simplexes cannot be visualized easily.

single factor. The one-dimensional simplex should have nonzero length: if the two vertexes coincide, the simplex will degenerate into a single point and lose its ability to move.

One-dimensional simplexes have been discussed by King and Deming [1]. An interesting discussion of one-dimensional movement is contained in the second part of the book *Flatland* by Edwin Abbott [2].

Two Dimensions

As we saw in Figure 3.1, a simplex in two dimensions must have three vertexes, and the geometric figure that has three vertexes is the triangle (Figure 3.2C). Contrary to what some writers have implied, two-dimensional simplexes do not have to be equilateral triangles, or isosceles triangles, or any other special kind of triangle. Two-dimensional simplexes should have some length and width so they do not degenerate into a one-dimensional line segment or a zero-dimensional point.

We will use two-dimensional simplexes to introduce the calculations and logic of the simplex algorithms. Later, the fundamental concepts will be extended to higher dimensional (and lower dimensional) factor spaces.

Three Dimensions

A simplex in three dimensions must have four vertexes. The geometric figure that has four corners is the tetrahedron (Figure 3.2D). Again, there are no special symmetry requirements for three-dimensional simplexes, but they should have some length, width, and depth so they do not degenerate into a two-dimensional quadrilateral or triangle, or a one-dimensional line segment, or a zero-dimensional point.

Higher Dimensions

A simplex in four or more dimensions must have five or more vertexes. Such higher dimensional simplexes are often referred to as "hypertetrahedra." These higher dimensional simplexes cannot be drawn, but they do exist and their properties are analogous to the properties of the simplexes that can be visualized.

Remember that in any number of dimensions, each corner of the simplex (each vertex) is used to represent a given set of experimental conditions in the factor space. From an experimental point of view, it is the vertexes that are the most important part of a simplex. It is the vertexes that describe how experiments should be carried out.

THE SIMPLEX REFLECTION MOVE

Figure 3.3 shows three examples of how a simplex can be moved into an adjacent area by rejecting one vertex (usually the vertex that gave the worst response) and projecting it through the average of the remaining vertexes to create one new vertex on the opposite side of the simplex [3]. This new vertex corresponds to a new set of experimental conditions that can then be evaluated.

In each example shown in Figure 3.3 the old, rejected vertex is connected to the retained vertexes by dashed lines. The vertexes joined by solid lines represent the new simplex. In Figures 3.3B and 3.3C, the small open circle shows the location of the average of the remaining vertexes. Because zero-dimensional simplexes have no freedom to move, they are not illustrated in Figure 3.3.

One Dimension

For a one-dimensional simplex (a line segment, Figure 3.3A), a new vertex is generated by rejecting one vertex and projecting it through the single remaining vertex an equal distance beyond. The new simplex is defined by the new vertex and the one retained old vertex. The new line segment is the same length as the old line segment.

Although it is tempting to view the generation of this new line segment as a simple translation (shifting) of the old line segment, such a conceptual view is misleading when extended to higher dimensional factor spaces. Thus, it is preferable to think of this move as a projection of one point (the old vertex) through a second point (the remaining vertex) an equal distance beyond to give the reflected point (the new vertex).

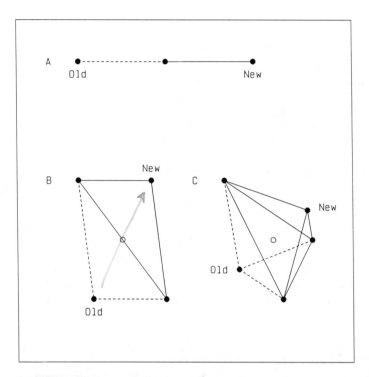

Figure 3.3 The simplex reflection move for (A) one-dimensional, (B) two-dimensional, and (C) three-dimensional factor spaces. Dashed lines represent the old simplex. Open circles show the average of the remaining vertexes.

Two Dimensions

For a two-dimensional simplex (a triangle, Figure 3.3B), a new vertex is generated by rejecting one vertex and projecting it through the *average* of the two remaining vertexes (that is, through the midpoint of one of the edges as shown by the open circle) an equal distance beyond. The new simplex is defined by the new vertex and the two retained old vertexes. The new triangle has the same area as the old triangle.

It is clear that, in general, the new triangle cannot be superimposed with the old triangle by simple translation (shifting). It is also clear that the new triangle is not a simple mirror image of the old triangle. In effect, the point-through-point reflection has turned the old simplex inside out to make the new simplex. Again, the simplex reflection move is a projection of one point (the old vertex) through a second point (the average of the remaining vertexes) an equal distance beyond to give the reflected point (the new vertex).

Three Dimensions

For a three-dimensional simplex (a tetrahedron, Figure 3.3C), a new vertex is generated by rejecting one vertex and projecting it through the average of the three remaining vertexes (that is, through the midpoint of one of the faces as shown by the open circle) an equal distance beyond. The new simplex is defined by the new vertex and the three retained old vertexes. The new tetrahedron has the same volume as the old tetrahedron.

Higher Dimensions

For a higher dimensional simplex (a hypertetrahedron), a new vertex is generated by rejecting one vertex and projecting it through the average of the k remaining vertexes (that is, through the midpoint of one of the hyperfaces) an equal distance beyond. The new simplex is defined by the new vertex and the k retained old vertexes. The new hypertetrahedron has the same hypervolume as the old hypertetrahedron.

Return from Hyperspace

The way simplexes move can be fascinating, and it is easy to get caught up in the mathematics and logic of the simplex moves. But three fundamental ideas should be remembered throughout:

- The simplex reflection is that of a point through a point. It is not a mirror-image reflection across a line (or across a plane, or across a hyperplane).
- The new vertex corresponds to a new set of experimental conditions (what statisticians call a "treatment" [4], or "treatment combination" [5], or "factor combination" [6], or "design point" [7]). Usually, the system under investigation must be evaluated at this set of conditions and a response must be obtained. Experiments *are* required!
- The purpose of the simplex is to move rapidly into the region of the optimum. The simplex can be very effective and efficient for this purpose. But when the simplex has located the region of the optimum, it then becomes relatively inefficient (although still effective) for finding the exact location of the optimum. It is usually best to quit using the simplex and switch over to a more powerful experimental design strategy for exploring the region of the optimum. More "classical" experimental designs such as three-level factorials, central-composite designs, Box–Behnken designs, and mixture designs are often useful for this purpose.

VECTOR REPRESENTATION OF VERTEXES

The coordinates of a vertex can be expressed concisely using mathematical vector notation. Each vertex is described by a row vector, an arrangement of numbers all written on one line and enclosed at each end by brackets or parentheses.

One Dimension

For a one-dimensional factor space, each vertex needs only one number to uniquely locate its position on the one-dimensional line. If the ith vertex is given the vector symbol V_i and the single dimension corresponds to the factor x_1, then the vertex is indicated symbolically as

$$V_i = [x_{1i}] \tag{3.1}$$

If $x_{1i} = 2.0$, for example, then the vertex is indicated numerically as

$$V_i = [2.0] \tag{3.2}$$

If the identity of the factor x_1 is pH, for example, then the experimental condition indicated by V_i corresponds to a pH of 2.0.

Two Dimensions

For a two-dimensional factor space, each vertex needs two numbers to uniquely locate its position in the two-dimensional plane. If the ith vertex is again given the vector symbol V_i and the two dimensions correspond to the factors x_1 and x_2, then the vertex is indicated symbolically as

$$V_i = [x_{1i}\ x_{2i}] \tag{3.3}$$

Although some writers would use V_i^T or V_i' to represent row vectors and reserve the notation V_i for column vectors only [8, 9], we will not make this distinction. If $x_{1i} = 3.7$ and $x_{2i} = 25.0$, for example, then the vertex is indicated numerically as

$$V_i = [3.7\ 25.0] \tag{3.4}$$

If the identity of the factor x_2 is Celsius temperature, for example, then the experimental conditions indicated by V_i correspond to a pH of 3.7 and a temperature of 25.0°C.

Three Dimensions

For a three-dimensional factor space, each vertex needs three numbers to uniquely locate its position in the three-dimensional volume. If the ith vertex is given the vector symbol V_i and the three dimensions correspond to the factors x_1, x_2, and x_3, then the vertex is indicated symbolically as

$$V_i = [x_{1i}\ x_{2i}\ x_{3i}] \tag{3.5}$$

If $x_{1i} = 4.2$, $x_{2i} = 37.0$, and $x_{3i} = 0.01$, for example, then the vertex is indicated numerically as

$$V_i = [4.2\ 37.0\ 0.01] \tag{3.6}$$

If the identity of the factor x_3 is the molar concentration of a particular reagent, for example, then the experimental conditions indicated by $\mathbf{V}_i$ correspond to a pH of 4.2, a temperature of 37.0°C, and a concentration of 0.01 M.

Higher Dimensions

For the general case of a vertex in k dimensions, k values must appear in brackets to uniquely locate the vertex in the k-dimensional factor space:

$$\mathbf{V}_i = [x_{1i}\, x_{2i}\, \cdots\, x_{ki}] \tag{3.7}$$

SIMPLEX TERMINOLOGY

- A **simplex** is a geometric figure defined by a number of points equal to one more than the number of dimensions in the factor space. The plural is either "simplexes" or "simplices" (SIM·pluh·sēs). We will use "simplexes."
- A **vertex** is a corner of a simplex. It is one of the points that define a simplex. It defines a set of experimental conditions. The plural is either "vertexes" or "vertices" (VER·tuh·sēs). We will use "vertexes."
- A **face** is that part of the simplex that remains after one of the vertexes is removed.
- A **hyperface** is the same as a face, but the term hyperface is usually used for four- and higher dimensional simplexes.
- A **centroid** is the center of mass of a set of vertex coordinates. There are two types of centroids that are of interest here:
  - The **simplex centroid** is the center of mass of all of the vertexes in the simplex.
  - The **centroid of the remaining hyperface** is the center of mass of the vertexes that remain when one vertex is removed from the simplex.

SIMPLEX CALCULATIONS

Let us assume that we want to use the sequential simplex algorithm to maximize (optimize) a response from a two-factor system. For purposes of illustration, we will simply list factor names and responses without specifying to what they refer. (If you would feel more comfortable with real factors, you might imagine that the factor x_1 corresponds to pH, the factor x_2 corresponds to temperature in °C, and the response y_1 is yield expressed in percent.)

Figure 3.4 shows three vertexes labeled **W** for the vertex giving the *worst* or *wastebasket* response (3.62), **B** for the vertex giving the *best* response (10.68), and **N** for the vertex giving the *next-to-the-worst* response (4.97).

(In two dimensions, vertex **N** is also the next-to-the-*best* vertex. However, in the case of more than two dimensions, the next-to-the-*best* vertex is not the same as the next-to-the-*worst* vertex. The designation **N** is reserved for the vertex giving the next-to-the-*worst* response.)

W = worst
B = Best
N = next-to-worst

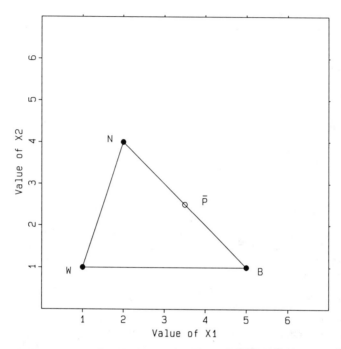

Figure 3.4 Initial simplex. **W** = vertex with the worst response; **N** = vertex with the next-to-the-worst response; **B** = vertex with the best response; **P̄** = centroid of the remaining hyperface.

The procedures for reflecting the simplex are very simple:

1. Initially, reject the vertex that has the worst response.
2. On all subsequent moves, reject the vertex that was labeled **N** in the previous simplex.

The following sections illustrate how the reflection move is actually carried out.

Calculation of the Centroid of the Remaining Hyperface

If the worst vertex **W** is removed from the simplex, the hyperface between **N** and **B** will remain. The average coordinates of these vertexes gives the location of the centroid of the remaining hyperface. This centroid is given the symbol **P̄**.

The coordinates of the centroid **P̄** are obtained by averaging the coordinates of the next-to-the-worst vertex **N** and the coordinates of the best vertex **B**, one factor at a time:

need a column for each factor

Simplex No. 1 → 2	Factor					
	X_1	X_2	Response	Rank	Vertex Number	Times Retained
Coordinates of	5.0	1.0	10.68	B	2	1
retained vertexes	2.0	4.0	4.97	N	3	1
Σ	7.0	5.0				
$\bar{P} = \Sigma/k$	3.5	2.5				
W					W	
$(\bar{P} - W)$						
$R = \bar{P} + (\bar{P} - W)$					R		0

# of factors

no responses for intermediate steps, only vertexes

Figure 3.5 Worksheet showing the calculation of $\bar{P}$, the centroid of the remaining hyperface, in two dimensions.

cont'd Fig 3.7 & further

	x_1	x_2
B	5.0	1.0
N	2.0	4.0
Σ	7.0	5.0
$\bar{P} = \Sigma/k$	3.5	2.5

In Figure 3.4, the coordinates of $\bar{P} = [3.5\ 2.5]$ are seen to represent the midpoint of the remaining hyperface.

Worksheets for Calculations

The previous calculation is more easily and systematically carried out on a worksheet [10]. The worksheet shown in Figure 3.5 contains information and calculations for generating the centroid of the hyperface discussed above. The various parts of the worksheet deserve some comment before we use it for further calculations. Blank worksheet forms that may be copied can be found in Appendix A.

Simplex No. In the upper left corner of the worksheet are two spaces to list simplex numbers. Because worksheets are used to go from one simplex to another, it is useful to fill these spaces with the number of the current simplex, followed by a right arrow, followed by the number of the next simplex. For example, the spaces in Figure 3.5 indicate that the current worksheet is being used to go from the first simplex to the second simplex: "Simplex No. *1 → 2.*"

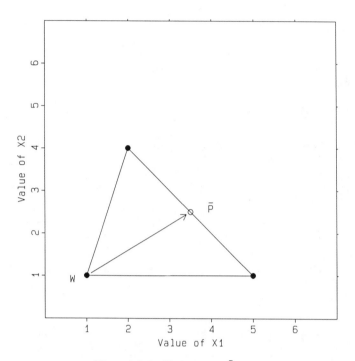

Figure 3.6 The vector ($\bar{\mathbf{P}}-\mathbf{W}$).

At the conclusion of each simplex move, information is transferred to a subsequent worksheet to begin the next move. That worksheet would then be labeled: "Simplex No. *2* → *3*." And so on.

Factor. To the right of the simplex numbers is a heading labeled "Factor," and under this heading are two columns labeled "x_1" and "x_2." These columns contain the x_1 and x_2 coordinates of the vertexes involved in the two-factor simplex. As we will see, all calculations on the worksheet are done in a column-wise fashion.

Response. To the right of the factors is a column labeled "Response." The responses are listed here. (Remember, you *do* have to carry out experiments!) Not every row is provided with a space for recording the response. Responses are listed only for the rows that correspond to simplex vertexes, not for rows containing intermediate calculations.

Rank. To the right of the responses is a column labeled "Rank." The letters in this column are **B**, the vertex giving the best response; **N**, normally the vertex giving the next-to-the-worst response; and **W**, normally the vertex giving the worst response, the vertex "in the wastebasket," the vertex that will be discarded.

When writing the initial simplex vertexes on the first worksheet, these absolute rankings must be followed. That is, the vertex giving the best response (**B**) must be listed on the top row. The vertex giving the next-to-the-worst response (**N**) must be listed on the bottom row of the section labeled "Coordinates of retained vertexes" (the second row in this case). And the worst vertex, the vertex to be rejected,

Simplex No. _1_ → _2_

	X₁	X₂	Response	Rank	Vertex Number	Times Retained
Coordinates of				B		
retained vertexes				N		
Σ						
$\bar{P} = \Sigma/k$	3.5	2.5				
W	1.0	1.0	3.62	W	1	1
$(\bar{P} - W)$	2.5	1.5				
$R = \bar{P} + (\bar{P} - W)$				R		0

Figure 3.7 Worksheet showing the calculation of $(\bar{P}-W)$.

the so-called wastebasket vertex (**W**), must be listed in the middle of the worksheet in the row provided for it.

Vertex Number. The column to the right of the rank is labeled "Vertex Number." This column provides a place to record each vertex's unchanging identity.

Each vertex for which a set of coordinates is calculated is given the next sequential vertex number. This assignment never changes and is retained as vertexes are transferred from one worksheet to the next. Even though a vertex might change its ranking, it never changes the vertex number assigned to it.

The numbering of vertexes in the initial simplex is usually arbitrary.

Times Retained. The column to the right of the vertex number is labeled "Times Retained." We will have more to say about this later, but for now simply treat the times retained column as a bookkeeping tool. When a vertex is transferred to the next worksheet, the corresponding value in the times retained column will be incremented by one.

Coordinates of Retained Vertexes. At the left side of the work sheet is the heading, "Coordinates of retained vertexes." The rows in this section simply contain information about the retained vertexes. All of these vertexes will eventually be transferred to the next worksheet.

Σ. The first row beneath the coordinates of retained vertexes is labeled with a capital Greek letter sigma (Σ), the mathematical symbol for summation. This row is used to sum the values above it as part of the calculation for the centroid of the remaining hyperface. The summations are carried out in a column-wise fashion. Thus, the first space on this row will contain the sum of the x_1 values for the best through the next-to-the-worst vertexes. The second space on this row will contain the sum of the x_2 values for the best through the next-to-the-worst vertexes. And so on, if there are more factors and therefore more columns.

$\bar{P} = \Sigma/k$. The next row is labeled $\bar{P}$ followed by the formula for calculating the centroid, the summation divided by the number of factors k.

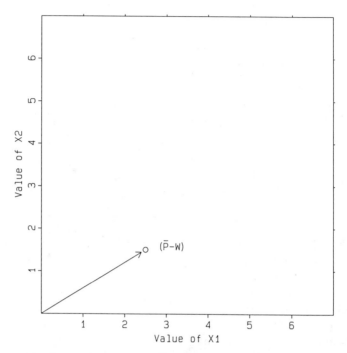

Figure 3.8 The vector ($\bar{\mathbf{P}}-\mathbf{W}$) translated to the origin.

W. The next row is provided for listing the coordinates of the rejected vertex, **W.** The coordinates of the rejected vertex are listed below the coordinates of the centroid to simplify the calculations that follow.

($\bar{\mathbf{P}} - \mathbf{W}$). This row will be discussed in more detail when we return to the calculations of the reflection move.

$\mathbf{R} = \bar{\mathbf{P}} + (\bar{\mathbf{P}} - \mathbf{W})$. Finally, a row labeled **R** for the reflection vertex is listed at the bottom of the worksheet. A formula for calculating **R** is also given.

Because the purpose of this worksheet is to find out where to do the next experiment, spaces are provided on this row (**R**) to record the response for this reflection vertex and to record its vertex number. The value in the times retained column is initialized to zero because this vertex has not yet appeared in any simplexes but the current one.

We return now to the calculations of the reflection move.

Calculation of ($\bar{P}$ – W)

Figure 3.6 illustrates the vector ($\bar{\mathbf{P}}-\mathbf{W}$). Although we could give this vector a simpler name (**Q**, for example), it has been found more descriptive to simply call it "p-bar minus w." This vector can be thought of as an arrow with its tail at **W** and its head at $\bar{\mathbf{P}}$. It has both length and direction.

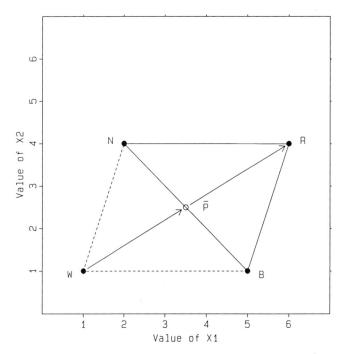

Figure 3.9   Generation of the vertex **R** by projection of **W** through **P̄** an equal distance beyond.

Note that $(\bar{\mathbf{P}}-\mathbf{W})$ is pointing in the direction we want to go if we are going to project **W** through **P̄** to produce the reflection vertex **R**. Not only is the vector $(\bar{\mathbf{P}}-\mathbf{W})$ pointing in the right direction, but it is also the right length for projecting an equal distance beyond **P̄** to establish the coordinates of the reflection vertex.

The calculation of $(\bar{\mathbf{P}}-\mathbf{W})$ is carried out as follows:

	x_1	x_2
P̄	3.5	2.5
W	1.0	1.0
$(\bar{\mathbf{P}}-\mathbf{W})$	2.5	1.5

This calculation is emphasized in the worksheet shown in Figure 3.7. By arranging the worksheet so **W** is directly under **P̄**, the subtraction to obtain $(\bar{\mathbf{P}}-\mathbf{W})$ is easily accomplished.

A geometric interpretation of the vector quantity $(\bar{\mathbf{P}}-\mathbf{W})$ is given in Figure 3.8, where the $(\bar{\mathbf{P}}-\mathbf{W})$ vector has been translated to position its tail at the origin. Its head is at the coordinates (2.5, 1.5). Note that it is still pointing in the same

Simplex No. _1_ → _2_	Factor		Response	Rank	Vertex Number	Times Retained
	x_1	x_2				
Coordinates of				B		
retained vertexes				N		
Σ						
$\bar{P} = Σ/k$	3.5	2.5				
W				W		
$(\bar{P} - W)$	2.5	1.5				
$R = \bar{P} + (\bar{P} - W)$	6.0	4.0		R	4	0

Figure 3.10 Worksheet showing the calculation of **R**.

direction it was in the previous figure. That is one of the nice things about vector quantities: you can move them around by adding or subtracting constant amounts, and they still point in the same direction.

Calculation of *R*

Figure 3.9 shows geometrically how the coordinates of the reflection vertex, **R**, are obtained by adding the vector $(\bar{P}-W)$ to the coordinates of $\bar{P}$. The effect is the same as if we had pushed $(\bar{P}-W)$ from its original location forward until the tail now coincides with $\bar{P}$. This is equivalent to projecting the coordinates of **W** through the coordinates of the centroid of the remaining hyperface $(\bar{P})$ an equal distance beyond.

The calculation of **R** is carried out as follows:

	x_1	x_2
$\bar{P} = Σ/k$	3.5	2.5
$(\bar{P}-W)$	2.5	1.5
R	6.0	4.0

Figure 3.10 shows the worksheet calculations for generating **R**. Adding the coordinates of $(\bar{P}-W)$ to the coordinates of $\bar{P}$ gives the coordinates of the reflection vertex **R**. The calculated numerical values can be verified geometrically in Figure 3.9. This reflection vertex is the fourth vertex to be encountered and is accordingly assigned vertex number 4.

COMPLETE WORKSHEET *(handwritten)*

(handwritten: current ↓, next ↙)

(handwritten: must be in bottom row of this section)

Simplex No. 1 → 2	Factor					
	x_1	x_2	Response	Rank	Vertex Number	Times Retained
Coordinates of	5.0	1.0	10.68	B	2	1
retained vertexes	2.0	4.0	4.97	N	3	1
Σ	7.0	5.0				
$\bar{P} = \Sigma/k$	3.5	2.5				
W	1.0	1.0	3.62	W	1	1
$(\bar{P} - W)$	2.5	1.5				
$R = \bar{P} + (\bar{P} - W)$	6.0	4.0	7.82	R	4	0

(handwritten: no response for these)

(handwritten: this row is discarded; don't next in worksheet)

Remember to evaluate the response
at the reflection vertex R

Figure 3.11 A complete simplex worksheet.

(handwritten: goes to "w" in next worksheet not this one)

Summary of Calculations

Figure 3.11 summarizes all of the previous individual calculations. Vertex #1 gave the worst response (**W**). Vertex #2 gave the best response (**B**). And vertex #3 gave the next-to-the-worst response (**N**). The initial simplex vertexes are always given a times-retained value of one. The reflection vertex **R** is vertex #4.

Recall that the purpose of the calculations on this worksheet was to determine the coordinates of **R**, coordinates that correspond to experimental values for x_1 and x_2. When an experiment was carried out at the calculated reflection vertex ($x_1 = 6.0$ and $x_2 = 4.0$), a response of 7.82 was obtained.

Moving to the Next Worksheet

So far, we have completed the operations required for the first simplex procedure:

- Initially, reject the vertex that has the worst response.

After this first procedure has been carried out on the initial simplex, all other moves are dictated by the second simplex procedure:

- On all subsequent moves, reject the vertex that was labeled **N** in the previous simplex.

This means transferring the contents of the row labeled **N** in the just-completed worksheet into the row labeled **W** on the next worksheet (see Figure 3.12).

Simplex No. _2_ → _3_	Factor		Response	Rank	Vertex Number	Times Retained
	X_1	X_2				
Coordinates of	5.0	1.0	10.68	B	2	2
retained vertexes	6.0	4.0	7.82	N	4	1
Σ						
$\bar{P}$ = Σ/k						
W	2.0	4.0	4.97	W	3	2
($\bar{P}$ − W)						
R = $\bar{P}$ + ($\bar{P}$ − W)				R		0

This row came from "N" row in previous table

Figure 3.12 Worksheet for the second simplex.

Usually (as in the example of vertex #3 in Figure 3.11), the next-to-the-worst vertex in the current simplex will turn out to be the worst vertex in the new simplex (see vertex #3 in Figure 3.12) and it would make sense to continue to call **W** the "worst vertex." However, as we will see later, sometimes the rejected vertex will *not* become the worst vertex. Thus, to avoid inconsistent terminology, in the second and all subsequent simplexes the vertex **W** will be called the "wastebasket vertex," the vertex that is being discarded.

Figure 3.12 shows the worksheet for the next simplex, the worksheet to be used in going from the second simplex to the third simplex. Compare this worksheet with the one shown in Figure 3.11.

The first thing to note is that vertex #1 in the initial simplex (see Figure 3.11) does not appear in the second simplex (see Figure 3.12). This is because vertex #1 was the worst vertex (**W**) in the initial simplex and was discarded. This is an obvious but important point to remember: once a vertex gets into the row labeled **W**, it never appears on the next worksheet, no matter what. It is "in the wastebasket" and gets thrown away. The contents of the row labeled **W** *never* get transferred to the next worksheet. You should eventually feel comfortable ignoring the contents of the row labeled **W** in the current worksheet as you move on to the next worksheet.

The second thing to note is that the next-to-the-worst vertex in the initial simplex (**N**, vertex #3 in Figure 3.11) is now the wastebasket vertex (**W**) in the second simplex (see Figure 3.12). This is simply a result of the second simplex rule: "on all subsequent moves, reject the vertex that was labeled **N** in the previous simplex." You should eventually feel comfortable automatically moving the contents of the row labeled **N** on the current worksheet into the row labeled **W** on the next worksheet.

Looking back at Figure 3.11, if **W** is gone forever and if **N** has been moved into the wastebasket row of the next worksheet, then the only remaining vertexes are rows **B** and **R**. These remaining vertexes must now be ranked and placed in their

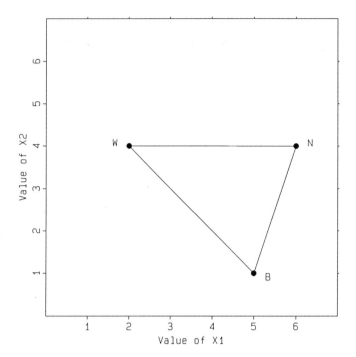

Figure 3.13 The second simplex. Vertexes labeled according to the worksheet shown in Figure 3.12.

proper rows in the section labeled "Coordinates of retained vertexes" on the next worksheet (Figure 3.12). In this example, the previous best vertex (**B**) is still the best vertex (**B**) in the new simplex. Thus, the reflection vertex (**R**) in the previous simplex becomes the next-to-the-worst vertex (**N**) in the new simplex.

In general, after the current **W** has been discarded and the vertex in the row labeled **N** has been transferred to the row labeled **W** on the next worksheet, the remaining retained vertexes are ranked and transferred to the next worksheet in order from best at the top, then next-to-the-best, and so on. This will fill in all of the rows at the top of the new worksheet.

As a vertex is transferred from a row of the previous worksheet to the current worksheet, the times retained column is incremented by one.

Figure 3.13 shows the new (second) simplex labeled according to the worksheet designations in Figure 3.12.

A summary of the rules for the basic simplex is given in Table 3.1.

A WORKED EXAMPLE

Assume it is desired to maximize the profit of a packaging machine. Two factors will be varied, throughput (e.g., items per minute) and amount of adhesive (e.g.,

Table 3.1 Rules for the Basic Simplex

1. Rank the vertexes of the first simplex on a worksheet in decreasing order of response from best to worst. Put the worst vertex into the row labeled **W**.
2. Calculate and evaluate **R**.
3. *Never* transfer the current row labeled **W** to the next worksheet. *Always* transfer the current row labeled **N** to the row labeled **W** on the next worksheet. Rank the remaining retained vertexes in order of decreasing response on the new worksheet, and go to 2.

milligrams per package). These will be called x_1 and x_2. The profit y_1 will be measured in dollars.

A starting simplex is established with the following vertex coordinates and responses:

Experiment	Throughput x_1	Adhesive x_2	Profit y_1
1	10.00	10.00	1.32
2	38.98	17.76	5.63
3	17.76	38.98	10.47

Simplex #1 → #2

Figure 3.14 shows a worksheet for carrying out the initial reflection move, "Simplex No. *1* → *2*." Figure 3.15 is a graph of the resulting operations. Vertex (experi-

Simplex No. 1 → 2	Factor					
	x_1	x_2	Response	Rank	Vertex Number	Times Retained
Coordinates of	17.76	38.98	10.47	B	3	1
retained vertexes	38.98	17.76	5.63	N	2	1
Σ	56.74	56.74				
$\bar{P} = Σ/k$	28.37	28.37				
W	10.00	10.00	1.32	W	1	1
$(\bar{P} - W)$	18.37	18.37				
R $= \bar{P} + (\bar{P} - W)$	46.74	46.74	88.02	R	4	0

Figure 3.14 First worksheet for the worked example.

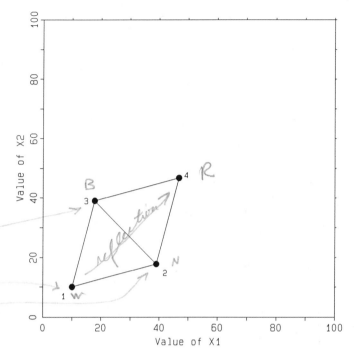

Figure 3.15   First reflection move for the worked example.

ment) #1 appears in the **W** row. Vertex #3 gives the best response and appears in the **B** row. Vertex #2 appears in the **N** row. The calculations of Σ, $\bar{\mathbf{P}}$, $(\bar{\mathbf{P}}-\mathbf{W})$, and **R** are shown. The reflection vertex **R** is located at a throughput of 46.74 and an amount of adhesive of 46.74.

When an experiment is carried out under the conditions of throughput and amount of adhesive suggested by this reflection vertex **R**, a profit of 88.02 is obtained. This is clearly better than the previous best vertex (10.47) and suggests that the simplex is moving in a favorable direction.

The vertex numbers listed on the worksheet (Figure 3.14) are "3" for **B**, "2" for **N**, and "1" for **W** in this example. The reflection vertex **R** becomes vertex #4. **Whenever the coordinates of a vertex are calculated, that vertex receives the next sequential vertex number.**

The times retained column contains the value "1" for each of the original vertexes and "0" for the new vertex **R**. These numbers will be incremented on going to the next worksheet.

The **W** vertex will *not* appear on the next worksheet. The wastebasket vertex **W** never appears on the next worksheet (Rule 3).

The vertexes that *will* appear on the next worksheet are **B**, **N**, and **R**. The vertex in the row labeled **N** is always transferred into the wastebasket row labeled **W** on the next worksheet (Rule 3). Of the remaining two transferred vertexes (**B** and **R** in the current simplex), the current **R** vertex is the better of the two and will

Simplex No. 2 → 3	Factor				Vertex Number	Times Retained
	X_1	X_2	Response	Rank		
Coordinates of	46.74	46.74	88.02	B	4	1
retained vertexes	17.76	38.98	10.47	N	3	2
Σ	64.50	85.72				
$\bar{P} = \Sigma/k$	32.25	42.86				
W	38.98	17.76	5.63	W	2	2
$(\bar{P} - W)$	-6.73	25.10				
$R = \bar{P} + (\bar{P} - W)$	25.52	67.96	71.14	R	5	0

Figure 3.16 Second worksheet for the worked example.

become the **B** vertex on the next worksheet. Thus, the current **B** vertex will become the **N** vertex on the next worksheet.

Figure 3.15 summarizes the reflection move.

Simplex #2 → #3

Figure 3.16 shows a worksheet for carrying out the next move, "Simplex No. *2 → 3.*" Figure 3.17 is a graph of the resulting operations. Vertex #2 appears in the **W** row. Vertex #4 gives the best response and appears in the **B** row. Vertex #3 appears in the **N** row. The reflection vertex **R** is located at a throughput of 25.52 and an amount of adhesive of 67.96.

Note that the quantity $(\bar{P}-W)$ is negative for x_1 in this example. This is entirely natural. It means that the x_1 component of $(\bar{P}-W)$ is directed toward lower values of x_1, not toward higher values as was the case in the previous move. When adding this quantity to $\bar{P}$ to generate **R**, it is important to remember that adding a negative quantity is the same as subtracting the absolute value of that quantity.

When an experiment is carried out under the conditions of throughput and amount of adhesive suggested by this reflection vertex **R**, a profit of 71.14 is obtained. This is not better than the best vertex **B** (88.02), but it is not worse than the vertex in the row labeled **N** (10.47).

The vertex numbers listed on the worksheet (Figure 3.16) are "4" for **B**, "3" for **N**, and "2" for **W** in this example. The reflection vertex **R** becomes vertex #5. Again, whenever the coordinates of a vertex are calculated, that vertex receives the next sequential vertex number.

The times retained column contains the value "1" for the best vertex **B**, "2" for the vertex in the row labeled **N**, and "2" for the wastebasket vertex **W**. These numbers were incremented from the values in the previous worksheet and will be incremented again on going to the next worksheet.

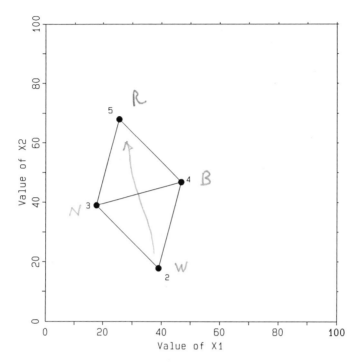

Figure 3.17 Second reflection move for the worked example.

Remember: the **W** vertex will *not* appear on the next worksheet. The wastebasket vertex **W** never appears on the next worksheet (Rule 3).

The vertexes that *will* appear on the next worksheet are **B**, **N**, and **R**. The vertex in the row labeled **N** is always transferred into the wastebasket row labeled **W** on the next worksheet (Rule 3). Of the remaining two transferred vertexes (**B** and **R** in the current simplex), the current **B** vertex is still the best vertex and will remain the **B** vertex on the next worksheet. Thus, the current **R** vertex will become the **N** vertex on the next worksheet.

Figure 3.17 summarizes the reflection move.

Simplex #3 → #4

Figure 3.18 shows a worksheet for carrying out the third reflection, "Simplex No. 3 → 4." Figure 3.19 is a graph of the resulting operations. Vertex #3 appears in the **W** row. Vertex #4 still gives the best response and appears in the **B** row. Vertex #5 appears in the **N** row. The reflection vertex **R** is located at a throughput of 54.50 and an amount of adhesive of 75.72.

When an experiment is carried out under the conditions of throughput and amount of adhesive suggested by this reflection vertex **R**, a profit of 43.97 is obtained. This is not better than the best vertex **B** (88.02). It is discouraging that

Simplex No. _3_ → _4_	Factor		Response	Rank	Vertex Number	Times Retained
	X_1	X_2				
Coordinates of	46.74	46.74	88.02	B	4	2
retained vertexes	25.52	67.96	71.14	N	5	1
Σ	72.26	114.70				
$\bar{P} = Σ/k$	36.13	57.35				
W	17.76	38.98	10.47	W	3	3
$(\bar{P} - W)$	18.37	18.37				
$R = \bar{P} + (\bar{P} - W)$	54.50	75.72	43.97	R	6	0

Figure 3.18 Third worksheet for the worked example.

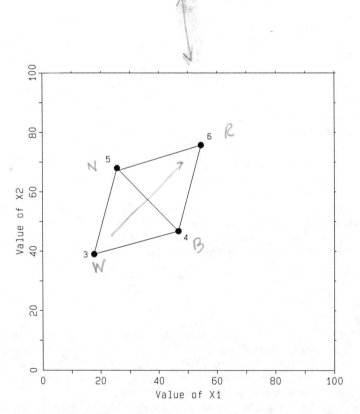

Figure 3.19 Third reflection move for the worked example.

Simplex No. _4_ → _5_	Factor		Response	Rank	Vertex Number	Times Retained
	X_1	X_2				
Coordinates of	46.74	46.74	88.02	B	4	3
retained vertexes	54.50	75.72	43.97	N	6	1
Σ	101.24	122.46				
$\bar{P}$ = Σ/k	50.62	61.23				
W	25.52	67.96	71.14	W	5	2
($\bar{P}$ − W)	25.10	−6.73				
R = $\bar{P}$ + ($\bar{P}$ − W)	75.72	54.50	81.49	R	7	0

worst vertex is not labelled "W" here

next page

Figure 3.20 Fourth worksheet for the worked example.

it is worse than the vertex in the row labeled **N** (71.14) but, having faith in the simplex, we continue on.

The vertex numbers listed on the worksheet (Figure 3.18) are "4" for **B**, "5" for **N**, and "3" for **W** in this example. The reflection vertex **R** becomes vertex #6. Remember: whenever the coordinates of a vertex are calculated, that vertex receives the next sequential vertex number.

The times retained column contains the value "2" for the best vertex **B**, "1" for the vertex in the row labeled **N**, and "3" for the wastebasket vertex **W**. These numbers were incremented from the values in the previous worksheet and will be incremented again on going to the next worksheet.

The wastebasket vertex **W** never appears on the next worksheet (Rule 3). The vertexes that appear on the next worksheet are **B**, **N**, and **R**. The vertex in the row labeled **N** is always transferred into the wastebasket row labeled **W** on the next worksheet (Rule 3). Of the remaining two transferred vertexes (**B** and **R** in the current simplex), the current **B** vertex is again the best vertex and will remain the **B** vertex on the next worksheet. The current **R** vertex will become the **N** vertex on the next worksheet.

Figure 3.19 summarizes this reflection move.

Simplex #4 → #5

Figure 3.20 shows a worksheet for carrying out the fourth reflection, "Simplex No. 4 → 5." Figure 3.21 is a graph of the resulting operations. Vertex #5 appears in the **W** row. Vertex #4 continues to give the best response and appears in the **B** row. Vertex #6 appears in the **N** row. The reflection vertex **R** is located at a throughput of 75.72 and an amount of adhesive of 54.50.

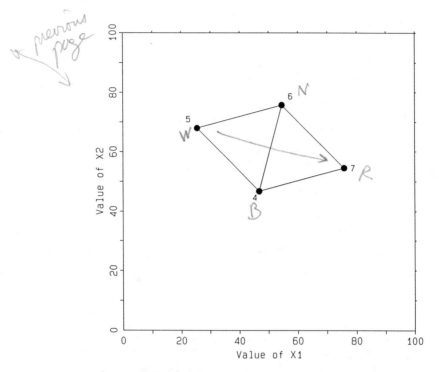

Figure 3.21 Fourth reflection move for the worked example.

Simplex No. _5_ → _6_

	X_1	X_2	Response	Rank	Vertex Number	Times Retained
Factor						
Coordinates of	46.74	46.74	88.02	B	4	4
retained vertexes	75.72	54.50	81.49	N	7	1
Σ	122.46	101.24				
$\bar{P} = \Sigma/k$	61.23	50.62				
W	54.50	75.72	43.97	W	6	2
$(\bar{P} - W)$	6.73	-25.10				
$R = \bar{P} + (\bar{P} - W)$	67.96	25.52	76.04	R	8	0

Figure 3.22 Fifth worksheet for the worked example.

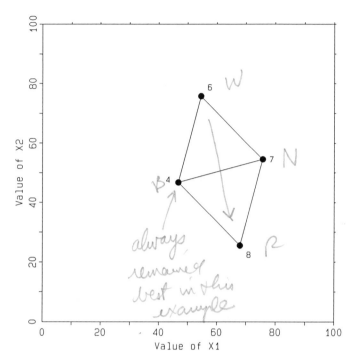

Figure 3.23 Fifth reflection move for the worked example.

Notice that on this worksheet the worst vertex is not in the row labeled **W**. The vertex that is in the row labeled **N** has a response that is worse than the wastebasket response **W**. This is why, after the first simplex, the vertex in the row labeled **W** is called the "wastebasket vertex," not necessarily the "worst vertex." (Similarly, the vertex **N** is not always the next-to-the-worst vertex, but we will usually still call it that.) This reversal happens occasionally with the simplex algorithms as they are presented in this text. It should not be a cause for concern. The algorithms are working properly.

When an experiment is carried out under the conditions of throughput and amount of adhesive suggested by this reflection vertex **R**, a profit of 81.49 is obtained. This is not better than the best vertex **B** (88.02), but it is better than vertex in the row labeled **N** (43.97).

The vertex numbers listed on the worksheet (Figure 3.20) are "4" for **B**, "6" for **N**, and "5" for **W** in this example. The reflection vertex **R** becomes vertex #7.

The times retained column contains the value "3" for the best vertex **B**, "1" for the vertex in the row labeled **N**, and "2" for the wastebasket vertex **W**.

The wastebasket vertex **W** never appears on the next worksheet (Rule 3). The vertexes that appear on the next worksheet are **B**, **N**, and **R**. The vertex in the row labeled **N** is always transferred into the wastebasket row labeled **W** on the next worksheet (Rule 3). Of the remaining two transferred vertexes (**B** and **R** in the

Simplex No. 6 → 7	Factor		Response	Rank	Vertex Number	Times Retained
	X_1	X_2				
Coordinates of	46.74	46.74	88.02	B	4	5
retained vertexes	67.96	25.52	76.04	N	8	1
Σ	114.70	72.26				
$\bar{P} = \Sigma/k$	57.35	36.13				
W	75.72	54.50	81.49	W	7	2
$(\bar{P} - W)$	-18.37	-18.37				
R $= \bar{P} + (\bar{P} - W)$	38.98	17.76	5.74	R	9	0

same as vertex 2

Figure 3.24 Sixth worksheet for the worked example.

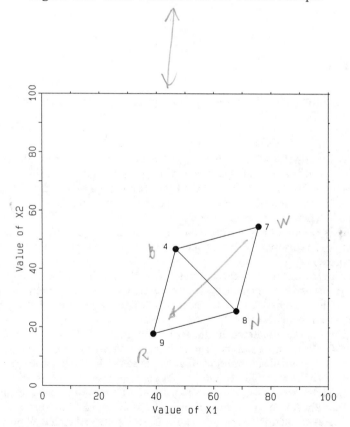

Figure 3.25 Sixth reflection move for the worked example.

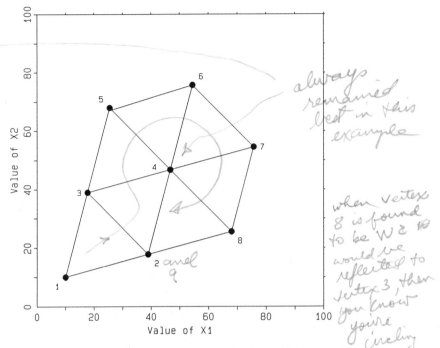

Figure 3.26 Summary of all simplex moves for the worked example.

[handwritten margin notes: "always remained best in this example"; "when vertex 8 is found to be W & it would be reflected to vertex 3, then you know you're circling"; "2 and 9"]

current simplex), the current **B** vertex continues to be the best vertex and will remain the **B** vertex on the next worksheet. The current **R** vertex will become the **N** vertex on the next worksheet.

Figure 3.21 summarizes the reflection move.

Simplex #5 → #6

Figure 3.22 shows a worksheet for carrying out the fifth reflection, "Simplex No. 5 → 6." Figure 3.23 is a graph of the resulting operations. Vertex #6 appears in the **W** row. Vertex #4 continues to give the best response and appears in the **B** row. Vertex #7 appears in the **N** row. The reflection vertex **R** is located at a throughput of 67.96 and an amount of adhesive of 25.52.

When an experiment is carried out under the conditions of throughput and amount of adhesive suggested by this reflection vertex **R**, a profit of 76.04 is obtained. This is still not better than the best vertex **B** (88.02).

Notice that the simplex has become fastened to vertex #4. Vertex #4 has been the best vertex in the last several simplexes. The simplexes are "spinning" around this very good vertex.

The vertex numbers listed on the worksheet (Figure 3.22) are "4" for **B**, "7" for **N**, and "6" for **W** in this example. The reflection vertex **R** becomes vertex #8.

Simplex No. ___ → ___	Factor		Response	Rank	Vertex Number	Times Retained
	X_1	X_2	Response	Rank	Vertex Number	Times Retained
Coordinates of				B		
retained vertexes				N		
Σ						
$\bar{P} = \Sigma/k$						
W				W		
$(\bar{P} - W)$						
$R = \bar{P} + (\bar{P} - W)$				R		0

Figure 3.27 Worksheet for two-factor basic simplex calculations.

The times retained column contains the value "4" for the best vertex **B**, "1" for the vertex in the row labeled **N**, and "2" for the wastebasket vertex **W**. Notice that the times retained value for vertex #4 is relatively large, confirming its continued presence in the simplexes.

The wastebasket vertex **W** never appears on the next worksheet (Rule 3). The vertex in the row labeled **N** is always transferred into the wastebasket row labeled **W** on the next worksheet (Rule 3). Of the remaining two transferred vertexes (**B** and **R** in the current simplex), the current **B** vertex continues to be the best vertex and will remain the **B** vertex on the next worksheet. The current **R** vertex will become the **N** vertex on the next worksheet.

Figure 3.23 summarizes the reflection move.

Simplex #6 → #7

Figure 3.24 shows a worksheet for carrying out the sixth reflection, "Simplex No. 6 → 7." Figure 3.25 is a graph of the resulting operations. Vertex #7 appears in the **W** row. Vertex #4 continues to give the best response and appears in the **B** row. Vertex #8 appears in the **N** row. The reflection vertex **R** is located at a throughput of 38.98 and an amount of adhesive of 17.76.

When an experiment is carried out under the conditions of throughput and amount of adhesive suggested by this reflection vertex **R**, a profit of only 5.74 is obtained.

The vertex numbers listed on the worksheet (Figure 3.24) are "4" for **B**, "8" for **N**, and "7" for **W** in this example. The reflection vertex **R** becomes vertex #9.

Notice that the coordinates of the new vertex (vertex #9) are the same as the coordinates of one of the earlier vertexes (vertex #2 in Figure 3.14). The two-

Table 3.2 Self-Test for the Basic Simplex Algorithm

Vertex	Factor x_1	Factor x_2	Response y_1
1. I	20.00	20.00	34.14
2. I	29.66	22.59	38.29
3. I	22.59	29.66	38.43
4. R	32.25	32.25	49.25
5. R	25.18	39.32	50.93
6. R	34.84	41.91	74.95
7. R	27.76	48.98	73.44
8. R	37.42	51.57	84.98
9. R	44.49	44.49	88.27
10. R	47.08	54.15	84.67
11. R	54.15	47.08	89.94
12. R	51.57	37.42	90.61
13. R	61.23	40.01	94.12
14. R	58.64	30.35	90.59
15. R	68.30	32.94	96.77
16. R	70.88	42.60	86.31
17. R	77.96	35.53	89.46
18. R	75.37	25.87	96.02
19. R	65.71	23.28	87.50
20. R	58.64	30.35	90.55
21. R	61.23	40.01	94.67
22. R	70.88	42.60	86.49

(Handwritten annotations: "Replicates" along vertices 16–20; "circling" with #13)

dimensional simplex has spun around completely and is beginning to overlap with vertexes that were carried out before.

This overlapping phenomenon is unique to one and two dimensions—straight line segments and triangles can close-pack and overlap in this way. Tetrahedra and hypertetrahedra do not close-pack and this overlapping phenomenon is not observed in three- and higher dimensional factor spaces.

It is not yet established that the simplexes are going to continue to spin around in a clockwise fashion and continue to repeat vertexes that have already been carried out. It is possible that the simplexes could become fastened to vertex #8 and begin to spin downward in a counterclockwise fashion. However, it is clear from the worksheet in Figure 3.24 that vertex #8 will be rejected on the next move to give a new reflection vertex that will overlap with vertex #3. Thus, it is now established that this two-dimensional simplex can make no further progress.

Figure 3.25 summarizes the reflection move.

Summary of the Basic Simplex Moves

Figure 3.26 summarizes the basic simplex moves. The initial simplex consists of vertexes 1–2–3. The simplex begins with a simple reflection. The second simplex

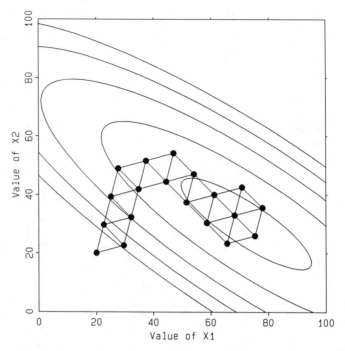

Figure 3.28 Summary of all simplex moves for the self-test for the basic simplex
algorithm.

contains the vertexes 2–3–4. The next move is a reflection to give the third simplex
containing the vertexes 3–4–5. The following moves are simple reflections that are
fastened about the best vertex #4. The simplex has circled when it is verified that
rejection of vertex #8 will produce a reflection that overlaps with vertex #3.

SELF-TEST FOR THE BASIC
SIMPLEX ALGORITHM

Make 19 copies of the two-factor simplex worksheet shown in Figure 3.27. Use
these copied worksheets to verify the calculations of the sequential simplex optimization summarized in Table 3.2. Verify the graphic results shown in Figure 3.28.
This optimization is discussed further in Chapter 5.

REFERENCES

1. P. G. King and S. N. Deming, "UNIPLEX: Single-factor optimization of
   response in the presence of error," *Anal. Chem.*, **46**(11), 1476–1481 (1974).
2. E. A. Abbott, *Flatland: A Romance of Many Dimensions*, 6th ed., Dover, New
   York, 1952, pp. 53–59.

3. W. Spendley, G. R. Hext, and F. R. Himsworth, "Sequential application of simplex designs in optimisation and evolutionary operation," *Technometrics*, **4**(4), 441–461 (1962).

4. R. A. Fisher, *Statistical Methods for Research Workers*, Hafner, New York, 1970.

5. G. E. P. Box, W. G. Hunter, and J. S. Hunter, *Statistics for Experimenters: An Introduction to Design, Data Analysis, and Model Building*, Wiley, New York, 1978.

6. S. N. Deming and S. L. Morgan, *Experimental Design: A Chemometric Approach*, Elsevier, Amsterdam, 1987.

7. A. I. Khuri and J. A. Cornell, *Response Surfaces: Designs and Analyses*, ASQC Quality Press, Milwaukee, WI, 1987.

8. G. H. Golub and C. F. Van Loan, *Matrix Computations*, 2nd ed., Johns Hopkins University Press, Baltimore, MD, 1989.

9. S. R. Searle, *Matrix Algebra Useful for Statistics*, Wiley, New York, 1987.

10. C. W. Lowe, "Some techniques of evolutionary operation," *Trans. Inst. Chem. Eng.*, **42**, T334–T344 (1964).

Chapter 4

The Variable-Size Simplex Algorithm

THE VARIABLE-SIZE SIMPLEX RULES

In 1965, Nelder and Mead [1] made two basic modifications to the original simplex algorithm of Spendley, Hext, and Himsworth [2]. The two modifications allow the simplex to expand in directions that are favorable and to contract in directions that are unfavorable. Because the modified algorithm allows the simplex to change its size, it is often referred to as the "variable-size simplex." In contrast, the original simplex algorithm is often called the "fixed-size simplex."

Although the logic behind the modifications might at first appear to be difficult, the additional moves are conceptually simple and can be understood easily. To aid in this understanding, we present one table and two figures that will be used to discuss the variable-size simplex algorithm.

Table 4.1 gives the written rules for the variable-size simplex algorithm. Note that Rules 1 and 3 for the variable-size algorithm are the same as Rules 1 and 3 for the fixed-size algorithm (see Table 3.1). It is only Rule 2 that is different. (In Table 4.1, the symbol ">" should be read, "is better than." The symbol "<" should be read, "is worse than." Similarly, the combination of symbols "$\geq$" should be read, "is better than or equal to." And the combination of symbols "$\leq$" should be read, "is worse than or equal to.")

Figure 4.1 illustrates the possible moves in the variable-size simplex algorithm. The familiar centroid of the remaining hyperface $\bar{P}$, the best vertex B, the next-to-the-worst vertex N, the worst vertex W, and the reflection vertex R are readily identified. New possibilities introduced here are the single expansion vertex E and two contraction vertexes: the contraction vertex on the reflection side C_R, and the contraction vertex on the worst side C_W.

Table 4.1 Rules for the Variable-Size Simplex

1. Rank the vertexes of the first simplex on a worksheet in decreasing order of response from best to worst. Put the worst vertex into the row labeled **W**.
2. Calculate and evaluate **R**:
   A. If **N** ≤ **R** ≤ **B**, use simplex **B..NR**, and go to 3.
   B. If **R** > **B**, calculate and evaluate **E**:
      i. If **E** ≥ **B**, use simplex **B..NE**, and go to 3.
      ii. If **E** < **B**, use simplex **B..NR**, and go to 3.
   C. If **R** < **N**:
      i. If **R** ≥ **W**, calculate and evaluate **C$_R$**, use simplex **B..NC$_R$**, and go to 3.
      ii. If **R** < **W**, calculate and evaluate **C$_W$**, use simplex **B..NC$_W$**, and go to 3.
3. *Never* transfer the current row labeled **W** to the next worksheet. *Always* transfer the current row labeled **N** to the row labeled **W** on the next worksheet. Rank the remaining retained vertexes in order of decreasing response on the new worksheet, and go to 2.

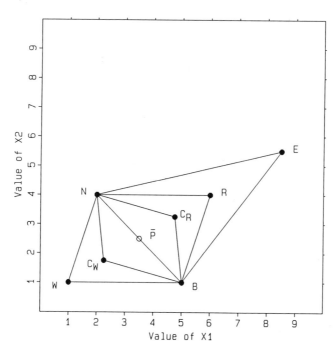

Figure 4.1 Possible moves in the variable-size simplex algorithm. $\bar{P}$ = centroid of the remaining hyperface; **B** = best vertex; **N** = next-to-the-worst vertex; **W** = worst (wastebasket) vertex; **R** = reflection vertex; **E** = expansion vertex; **C$_R$** = contraction vertex on the **R** side; **C$_W$** = contraction vertex on the **W** side.

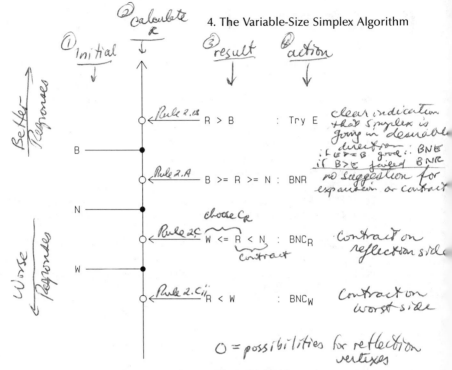

Figure 4.2 contains handwritten annotations including: "calculate R", "initial", "result", "action", "Better Responses", "Worse Responses", and to the right of the response line various notes such as "clear indication that simplex is going in desirable direction", "if R = B good: BNE", "if B > E failed BNR", "no suggestion for expansion or contract", "Contract on reflection side", "Contract on worst side", "O = possibilities for reflection vertexes".

The response line shows:
- Rule 2.B R > B : Try E
- B
- Rule 2.A B >= R >= N : BNR
- N
- choose C_R Rule 2C W <= R < N : BNC_R (contract)
- W
- Rule 2.C'' R < W : BNC_W

Figure 4.2 Logic of the possible moves in the variable-size simplex algorithm.

Figure 4.2 contains a vertical line of response that is used to show the logic of the possible variable-size simplex moves. Better responses lie toward the top of the line; worse responses lie toward the bottom of the line. The responses of the best vertex **B**, the next-to-the-worst vertex **N**, and the worst vertex **W** are identified to the left of the response line. The four possibilities for the response of the reflection vertex **R** are identified to the right of the response line.

Table 4.1, Figure 4.1, and Figure 4.2 all contain similar information presented from different points of view. All three are used to discuss the variable-size simplex rules.

Rule 1

The first major rule for the variable-size simplex (Table 4.1) is the same as the first rule for the fixed-size simplex (Table 3.1):

> Rank the vertexes of the first simplex on a worksheet in decreasing order of response from best to worst. Put the worst vertex into the row labeled **W**.

Thus, nothing new is introduced by this first rule.

Rule 3

The third major rule for the variable-size simplex (Table 4.1) is the same as the third rule for the fixed-size simplex (Table 3.1):

Never transfer the current row labeled **W** to the next worksheet. *Always* transfer the current row labeled **N** to the row labeled **W** on the next worksheet. Rank the remaining retained vertexes in order of decreasing response on the new worksheet, and go to [Rule] 2.

Thus, nothing new is introduced by this third rule, either.

Rule 2

The second major rule for the variable-size simplex (Table 4.1) starts out the same as the second rule for the fixed-size simplex (Table 3.1):

Calculate and evaluate **R**.

Up to this point, nothing new is introduced by this rule. The calculations for the reflection vertex are identical to those developed in Chapter 3: $\mathbf{R} = \bar{\mathbf{P}} + (\bar{\mathbf{P}} - \mathbf{W})$. These calculations were illustrated in Figures 3.4 through 3.13.

With the variable-size simplex, four possible outcomes are now considered for the response at the reflection vertex **R**. Only one of the four possibilities can occur on any given simplex move. These possibilities (and the five simplex moves that result) are discussed in turn.

Rule 2.A: If N ≦ R ≦ B, Use Simplex B..NR

If the response at the reflection vertex **R** is not better than the response at the previously best vertex **B**, then there is no clear evidence that the simplex is moving in a particularly desirable direction. And if the response at the reflection vertex **R** is not worse than the response at the next-to-the-worst vertex **N**, then there is no clear evidence that the simplex is moving in a particularly undesirable direction, either.

This is what is called informally a "ho-hum" vertex (yawn). It is routine. It is unexciting. It is dull. It is not better than the best. It is not worse than the next-to-the-worst. There is nothing to suggest that the simplex should be expanded. There is nothing to suggest that the simplex should be contracted. In this case, no additional moves are called for in the current simplex, and the algorithm behaves like the fixed-size algorithm on this move. The new simplex will consist of the current vertexes **B..NR**.

Rule 2.B: If R > B, Calculate and Evaluate E

If the response at **R** is better than the response at **B**, then this is a clear indication the simplex is progressing in a desirable direction. It makes sense to try an experiment farther out along the projection of $(\bar{\mathbf{P}} - \mathbf{W})$ in the direction of apparently improving response. Nelder and Mead [1] suggested going out twice as far from $\bar{\mathbf{P}}$ as the reflection vertex **R**. This new vertex is labeled **E**, for an "expansion vertex": $\mathbf{E} = \bar{\mathbf{P}} + 2(\bar{\mathbf{P}} - \mathbf{W}) = \mathbf{R} + (\bar{\mathbf{P}} - \mathbf{W})$.

One of two outcomes will occur when the vertex **E** is evaluated.

Rule 2.B.i: If $E \geq B$, Use Simplex $B..NE$

If E is better than or equal to B, then the expansion is considered to be successful. The reflection vertex R is disregarded, and the new simplex will consist of the current vertexes $B..NE$. A successful expansion causes the simplex to become elongated in a favorable direction.

Rule 2.B.ii: If $E < B$, Use Simplex $B..NR$

If E is worse than B, then the expansion is considered to have failed. This can occur, for example, if the reflection approaches an optimum and the expansion extends the simplex beyond the optimum toward less desirable responses (i.e., if E takes the simplex "over the hill"). If this occurs, the expansion vertex E is disregarded, and the new simplex will consist of the current vertexes $B..NR$. A failed expansion causes the simplex to retain its current size.

Rule 2.C: If $R < N$, Contract

If the response at the reflection vertex R is worse than the response at the next-to-the-worst vertex N, this is a clear indication that the simplex is moving in an undesirable direction and should contract so it does not continue to make relatively large moves in this direction.

There are two possible contractions. The simplex could be contracted on the R side of $\bar{P}$, a so-called C_R contraction. Or the simplex could be contracted on the W side of $\bar{P}$, a so-called C_W contraction. Which of these two possibilities is chosen depends on just how bad the response actually is at the reflection vertex R.

Sometimes in life a choice is simply between one good thing and one bad thing. At other times, the choice is more difficult because it is between two bad things. When such decisions must be made, it is useful to follow the philosophy of "choosing the lesser of the evils." Such a situation exists when the variable-size simplex algorithm has to choose between the C_R and C_W contractions. Both of them will probably give a poor response, but the trick is to choose the contraction that is more likely to give the better of the two.

Contractions are handled differently from expansions because contractions are not tested to see if they should be kept. If a contraction is carried out, it is retained no matter what the response might be. Expansions, it will be recalled, have to be tested to see if they will be retained.

Rule 2.C.i: If $R \geq W$, Use Simplex $B..NC_R$

If the response at the reflection vertex R is better than or equal to the response at the rejected vertex W, then the R side of $\bar{P}$ looks better than (or as good as) the W side of $\bar{P}$, and the C_R contraction should be chosen in this case. The C_R contraction vertex lies halfway between $\bar{P}$ and R. Thus, the calculation for the contraction on the reflection side is given by $C_R = \bar{P} + \frac{1}{2}(\bar{P}-W)$. The new simplex will consist of the vertexes $B..NC_R$.

Simplex No. ___ → ___	Factor					
	X_1	X_2	Response	Rank	Vertex Number	Times Retained
Coordinates of				B		
retained vertexes				N		
Σ						
$\bar{P} = \Sigma/k$						
W				W		
$(\bar{P} - W)$						
$R = \bar{P} + (\bar{P} - W)$				R		0
$(\bar{P} - W)/2$			← *used for either contraction*			
$C_W = \bar{P} - (\bar{P} - W)/2$				C_W		0
$C_r = \bar{P} + (\bar{P} - W)/2$				C_r		0
$E = R + (\bar{P} - W)$				E		0

Figure 4.3 Worksheet for two-factor variable-size simplex calculations.

Rule 2.C.ii: If $R < W$, Use Simplex $B..NC_W$

If the response at the reflection vertex $\mathbf{R}$ is worse than the response at the rejected vertex $\mathbf{W}$, then the $\mathbf{W}$ side of $\bar{\mathbf{P}}$ looks better than the $\mathbf{R}$ side of $\bar{\mathbf{P}}$, and the C_W contraction should be chosen in this case. The C_W contraction vertex lies halfway between $\bar{\mathbf{P}}$ and $\mathbf{W}$. Thus, the calculation for the contraction on the worst side is given by $C_W = \bar{\mathbf{P}} - \frac{1}{2}(\bar{\mathbf{P}} - \mathbf{W})$. The new simplex will consist of the vertexes $B..NC_W$.

THE VARIABLE-SIZE WORKSHEET

Figure 4.3 is a blank variable-size worksheet. Most of it is identical to the fixed-size worksheet shown in Figure 3.5. However, the variable-size worksheet has four additional rows at the bottom. It is possible that on any given simplex move, none of these rows will be used (Rule 2.A). At most, only two of these rows will be used (Rule 2.C).

$(\bar{P}-W)/2$ Row

The fourth row from the bottom is for the calculation of the quantity $\frac{1}{2}(\bar{P}-W)$, a quantity that is used when either contraction is carried out. It is obtained by multiplying the quantity $(\bar{P}-W)$ two rows above by 0.5. Unless a contraction is to be carried out, this row should not be used.

$C_W = \bar{P} - (\bar{P}-W)/2$ Row

This row is used for calculating the coordinates of a C_W contraction. The quantity $\frac{1}{2}(\bar{P}-W)$ is on the row immediately above; the quantity $\bar{P}$ is four rows above that. A simple subtraction of the two rows produces the coordinates of the C_W contraction.

$C_R = \bar{P} + (\bar{P}-W)/2$ Row

This row is used for calculating the coordinates of a C_R contraction. The quantity $\frac{1}{2}(\bar{P}-W)$ is two rows above; the quantity $\bar{P}$ is four rows above that. A simple addition of the two rows produces the coordinates of the C_R contraction.

$E = R + (\bar{P}-W)$ Row

This row is used for calculating the coordinates of an expansion vertex $\mathbf{E}$. The quantity $\mathbf{R}$ is four rows above; the quantity $(\bar{P}-W)$ is on the row above that. A simple addition of the two rows produces the coordinates of the expansion $\mathbf{E}$.

A WORKED EXAMPLE

For purposes of illustration, assume it is desired to maximize the percentage yield of a chemical reaction. Two factors will be varied, temperature and pressure. These will be called x_1 and x_2. A starting simplex is established with the following vertex coordinates and responses:

Experiment	Temperature	Pressure	% Yield
1	15.00	15.00	8.46
2	34.32	20.18	36.19
3	20.18	34.32	42.37

Reflection and Retained Expansion

Figure 4.4 shows a worksheet for carrying out the initial reflection move, "Simplex No. $1 \rightarrow 2$." Figure 4.5 is a graph of the resulting operations. Vertex (experiment) #1 gives the worst response and appears in the $\mathbf{W}$ row. Vertex #3 gives the best response and appears in the $\mathbf{B}$ row. Vertex #2 gives the next-to-the-worst response

Simplex No. _1_ → _2_

	X_1	X_2	Response	Rank	Vertex Number	Times Retained
Coordinates of	20.18	34.32	42.37	B	3	1
retained vertexes	34.32	20.18	36.19	N	2	1
Σ	54.50	54.50				
$\bar{P} = \Sigma/k$	27.25	27.25				
W	15.00	15.00	8.46	W	1	1
$(\bar{P} - W)$	12.25	12.25				
$R = \bar{P} + (\bar{P} - W)$	39.50	39.50	65.87	R	4	0
$(\bar{P} - W)/2$						
$C_W = \bar{P} - (\bar{P} - W)/2$				C_W		0
$C_r = \bar{P} + (\bar{P} - W)/2$				C_r		0
$E = R + (\bar{P} - W)$	51.75	51.75	81.76	E	5	0

Figure 4.4 Worksheet showing reflection followed by retained expansion.

and appears in the **N** row. The calculations of Σ, $\bar{P}$, $(\bar{P}-W)$, and **R** are familiar and follow the procedures established for the fixed-size simplex. The reflection vertex **R** is located at a temperature of 39.50 and a pressure of 39.50.

When an experiment is carried out under the conditions of temperature and pressure suggested by this reflection vertex **R**, a yield of 65.87% is obtained. This is clearly better than the previous best vertex (42.37%) and suggests that the simplex is moving in a favorable direction. Thus, an expansion is attempted.

On the worksheet in Figure 4.4, the expansion is formed by adding another $(\bar{P}-W)$ to **R**. The expansion vertex **E** is located at a temperature of 51.75 and a pressure of 51.75.

When an experiment is carried out under the conditions of temperature and pressure suggested by this expansion vertex, a yield of 81.76% is obtained. Thus, the response at **E** is better than the response at **B** (42.37%), and the expansion vertex is retained.

Note that the second, third, and fourth rows from the bottom of the worksheet were not used. These rows are used for contractions only.

The vertex numbers listed on the worksheet (Figure 4.4) are "3" for **B**, "2" for **N**, and "1" for **W** in this example. The reflection vertex **R** becomes vertex #4. The expansion vertex **E** becomes vertex #5. **Whenever the coordinates of a vertex are calculated, that vertex receives the next sequential vertex number, whether it is retained in the next simplex or not.**

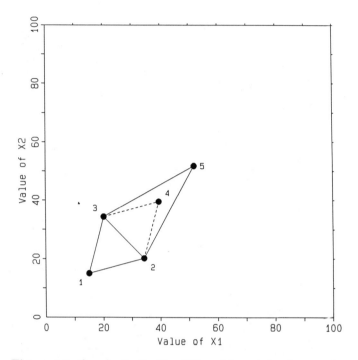

Figure 4.5 Graph of reflection followed by retained expansion.

The times retained column contains the value "1" for each of the original vertexes, and "0" for each of the new vertexes (**R** and **E**). These numbers will be incremented on going to the next worksheet.

The vertexes that will *not* appear on the next worksheet are **W** and **R**. As with the fixed-size simplex, the wastebasket vertex **W** never appears on the next worksheet (Rule 3). With the variable-size simplex, when an expansion has been carried out either **R** or **E** but not both can be transferred to the next worksheet. Because the expansion was successful here (**E** ≥ **B**), **R** is *not* transferred to the next worksheet.

The vertexes that *will* appear on the next worksheet are **B**, **N**, and **E**. As with the fixed-size simplex, the next-to-the-worst vertex **N** is always transferred into the wastebasket row labeled **W** on the next worksheet (Rule 3). Of the remaining two transferred vertexes (**B** and **E** in the current simplex), the current **E** vertex is the better of the two and will become the **B** vertex on the next worksheet. Thus, the current **B** vertex will become the **N** vertex on the next worksheet.

Figure 4.5 summarizes the reflection and the retained expansion. It is a convention that nonretained vertexes are connected by dashed lines; retained vertexes are connected by solid lines. Thus, vertex #4 (the initial reflection **R**) will not be retained and is connected to the original simplex with dashed lines. Vertex #5 (the expansion **E**) will be retained and is connected to the original simplex with solid lines to form the second complete simplex.

Simplex No. _2_ → _3_	Factor		Response	Rank	Vertex Number	Times Retained
	X_1	X_2				
Coordinates of	51.75	51.75	81.76	B	5	1
retained vertexes	20.18	34.32	42.37	N	3	2
Σ	71.93	86.07				
$\bar{P} = \Sigma/k$	35.96	43.04				
W	34.32	20.18	36.19	W	2	2
$(\bar{P} - W)$	1.64	22.86				
$R = \bar{P} + (\bar{P} - W)$	37.60	65.90	83.11	R	6	0
$(\bar{P} - W)/2$						
$C_W = \bar{P} - (\bar{P} - W)/2$				C_W		0
$C_r = \bar{P} + (\bar{P} - W)/2$				C_r		0
$E = R + (\bar{P} - W)$	39.24	88.76	62.48	E	7	0

Figure 4.6 Worksheet showing reflection followed by nonretained expansion.

Reflection and Nonretained Expansion

Figure 4.6 shows a worksheet for carrying out the next simplex move, "Simplex No. $2 \rightarrow 3$." Figure 4.7 is a graph of the resulting operations. Vertex #2 is the wastebasket vertex and appears in the **W** row. Vertex #5 gives the best response and appears in the **B** row. Vertex #3 gives the next-to-the-worst response and appears in the **N** row. The reflection vertex **R** is located at a temperature of 37.60 and a pressure of 65.90.

When an experiment is carried out under the conditions of temperature and pressure suggested by this reflection vertex **R**, a yield of 83.11% is obtained. This is better than the previous best vertex (81.76%) and suggests that the simplex is still moving in a favorable direction. Thus, another expansion is attempted.

When an experiment is carried out under the conditions of temperature and pressure suggested by this expansion vertex, a yield of 62.48% is obtained. This response at **E** is not better than the response at **B** (81.76%), and the expansion vertex will not be retained. It appears that the simplex has overshot the optimal region.

The vertex numbers listed on the worksheet (Figure 4.6) are "5" for **B**, "3" for **N**, and "2" for **W**. The reflection vertex **R** is vertex #6. The expansion vertex **E** becomes vertex #7. Again, whenever the coordinates of a vertex are calculated, that

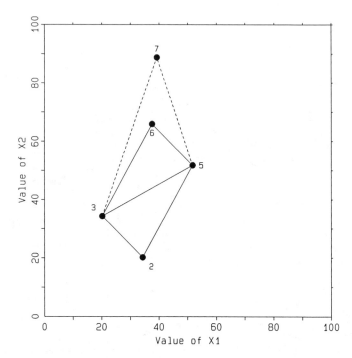

Figure 4.7   Graph of reflection followed by nonretained expansion.

vertex receives the next sequential vertex number, whether it is retained in the next simplex or not.

The times retained column contains the value "1" for the best vertex **B**, and "2" for the vertexes **N** and **W**. The new vertexes **R** and **E** are each given a times retained value of "0."

In this current simplex, the vertexes that will *not* appear on the next worksheet are **W** and **E**. The wastebasket vertex **W** never appears on the next worksheet (Rule 3). Because the expansion was *not* successful here (**E** < **B**), **E** is *not* transferred to the next worksheet.

The vertexes that *will* appear on the next worksheet are **B**, **N**, and **R**. Again, the next-to-the-worst vertex **N** is always transferred into the wastebasket row labeled **W** on the next worksheet (Rule 3). Of the remaining two transferred vertexes (**B** and **R** in the current simplex), the current **R** vertex is the better of the two and will become the **B** vertex on the next worksheet. Thus, the current **B** vertex will become the **N** vertex on the next worksheet.

Figure 4.7 summarizes the reflection and the nonretained expansion. Vertex #6 (the reflection **R**) will be retained and is connected to the original simplex with solid lines to form the third complete simplex. Vertex #7 (the expansion **E**) will not be retained and is connected to the original simplex with dashed lines.

Simplex No. _3_ → _4_	Factor		Response	Rank	Vertex Number	Times Retained
	X_1	X_2				
Coordinates of	37.60	65.90	83.11	B	6	1
retained vertexes	51.75	51.75	81.76	N	5	2
Σ	89.35	117.65				
$\bar{P} = \Sigma/k$	44.68	58.82				
W	20.18	34.32	42.37	W	3	3
$(\bar{P} - W)$	24.50	24.50				
$R = \bar{P} + (\bar{P} - W)$	69.18	83.32	75.31	R	8	0
$(\bar{P} - W)/2$	12.25	12.25				
$C_W = \bar{P} - (\bar{P} - W)/2$				C_W		0
$C_r = \bar{P} + (\bar{P} - W)/2$	56.93	71.07	84.62	C_r	9	0
$E = R + (\bar{P} - W)$				E		0

Figure 4.8 Worksheet showing reflection followed by contraction on the reflection side.

Contraction on the Reflection Side

Figure 4.8 shows a worksheet for carrying out the next simplex move, "Simplex No. *3 → 4*." Figure 4.9 is a graph of the resulting operations. Vertex #3 is the wastebasket vertex and appears in the **W** row. Vertex #6 gives the best response and appears in the top **B** row. Vertex #5 gives the next-to-the-worst response and appears in the **N** row. The reflection vertex **R** is located at a temperature of 69.18 and a pressure of 83.32.

When an experiment is carried out under the conditions of temperature and pressure suggested by this reflection vertex **R**, a yield of 75.31% is obtained. This is not better than the previous best vertex (83.11%), so an expansion is not suggested. The response at **R** is, in fact, worse than the response at **N**, which suggests that the simplex is moving in an unfavorable direction. Thus, a contraction is indicated (see Table 4.1 and Figure 4.2).

At this point a decision must be made as to which contraction to carry out: C_R or C_W. If **R** were worse than **W**, then a C_W contraction would be indicated (see Table 4.1 and Figure 4.2). Instead, **R** is worse than **N** but not worse than **W**, so a C_R contraction will be carried out. In short, although the response at the reflection vertex **R** is not very good, it is not as bad as the response on the other side

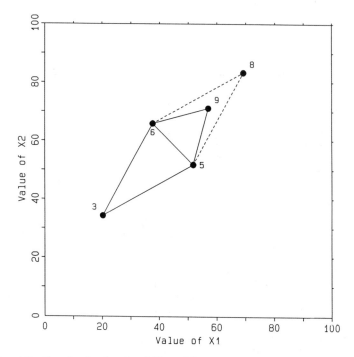

Figure 4.9 Graph of reflection followed by contraction on the reflection side.

of the remaining hyperface (i.e., on the **W** side), so a contraction is made on the reflection side.

When an experiment is carried out under the conditions of temperature and pressure suggested by this contraction vertex, a yield of 84.62% is obtained. This is a very good yield, better than the best. But as far as the variable-size simplex algorithm is concerned, it is irrelevant how good the response is at this C_R vertex. Contractions are not tested to see if they should be kept. If a contraction is carried out, it is retained no matter what the response might be.

The vertex numbers listed on the worksheet (Figure 4.8) are "6" for **B**, "5" for **N**, and "3" for **W**. The reflection vertex **R** is vertex #8. The contraction vertex C_R becomes vertex #9. Whenever the coordinates of a vertex are calculated, that vertex receives the next sequential vertex number, whether it is retained in the next simplex or not.

The times retained column contains the value "1" for the best vertex **B**, "2" for the next-to-the-worst vertex **N**, and "3" for the wastebasket vertex **W**. The new vertexes **R** and C_R are each given a times retained value of "0."

The vertexes that will *not* appear on the next worksheet are **W** and **R**. The wastebasket vertex **W** never appears on the next worksheet (Rule 3). If a contraction is carried out, the reflection vertex **R** is never carried to the next worksheet.

The vertexes that *will* appear on the next worksheet are **B**, **N**, and C_R. The next-to-the-worst vertex **N** is always transferred into the wastebasket row labeled

Simplex No. _4_ → _5_	Factor					
	x_1	x_2	Response	Rank	Vertex Number	Times Retained
Coordinates of	56.93	71.07	84.62	B	9	1
retained vertexes	37.60	65.90	83.11	N	6	2
Σ	94.53	136.97				
$\bar{P}$ = Σ/k	47.26	68.48				
W	51.75	51.75	81.76	W	5	3
$(\bar{P} - W)$	-4.49	16.73				
R = $\bar{P}$ + $(\bar{P} - W)$	42.77	85.21	67.17	R	10	0
$(\bar{P} - W)/2$	-2.24	8.36				
C_W = $\bar{P}$ - $(\bar{P} - W)/2$	49.50	60.12	82.68	C_W	11	0
C_r = $\bar{P}$ + $(\bar{P} - W)/2$				C_r		0
E = R + $(\bar{P} - W)$				E		0

Figure 4.10 Worksheet showing reflection followed by contraction on the worst (wastebasket) side.

W on the next worksheet (Rule 3). Of the remaining two transferred vertexes (**B** and C_R in the current simplex), the current C_R vertex is the better of the two and will become the **B** vertex on the next worksheet. Thus, the current **B** vertex will become the **N** vertex on the next worksheet.

Figure 4.9 summarizes the reflection followed by the C_R contraction. Vertex #8 (the reflection **R**) will not be retained and is connected to the original simplex with dashed lines. Vertex #9 (the contraction C_R) will be retained and is connected to the original simplex with solid lines to form the fourth complete simplex.

Contraction on the Worst Side

Figure 4.10 shows a worksheet for carrying out the next simplex move, "Simplex No. 4 → 5." Figure 4.11 is a graph of the resulting operations. Vertex #5 is now the wastebasket vertex and appears in the **W** row. Vertex #9 gives the best response and appears in the top **B** row. Vertex #6 gives the next-to-the-worst response and appears in the **N** row. The reflection vertex **R** is located at a temperature of 42.77 and a pressure of 85.21.

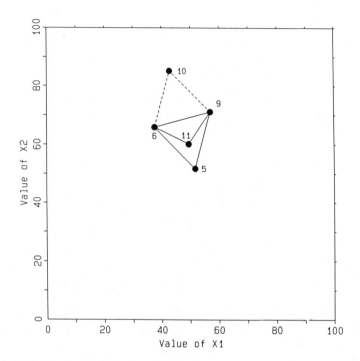

Figure 4.11 Graph of reflection followed by contraction on the worst (waste-basket) side.

Note that the quantity $(\bar{\mathbf{P}}-\mathbf{W})$ is negative for x_1 in this example. This is entirely natural. It means that the x_1 component of $(\bar{\mathbf{P}}-\mathbf{W})$ is directed toward lower values of x_1, not toward higher values as has been the case up to this point. When adding this quantity to $\bar{\mathbf{P}}$ to generate $\mathbf{R}$, it is important to remember that adding a negative quantity is the same as subtracting the absolute value of that quantity.

When an experiment is carried out under the conditions of temperature and pressure suggested by this reflection vertex $\mathbf{R}$, a yield of 67.17% is obtained. The response at $\mathbf{R}$ is again worse than the response at $\mathbf{N}$, which suggests that the simplex is moving in an unfavorable direction. Thus, another contraction is indicated.

This time, $\mathbf{R}$ is worse than $\mathbf{W}$, and a $\mathbf{C_W}$ contraction is indicated (see Table 4.1 and Figure 4.2). This is clearly the correct choice: the response at the reflection vertex $\mathbf{R}$ is worse than the response on the other side of the remaining hyperface (i.e., on the $\mathbf{W}$ side), so a contraction is made back to the $\mathbf{W}$ side.

Note that the quantity $(\bar{\mathbf{P}}-\mathbf{W})/2$ is negative for x_1 in this example. The calculation $\mathbf{C_W} = \bar{\mathbf{P}} - (\bar{\mathbf{P}}-\mathbf{W})/2$ now involves subtracting a negative quantity, which is the same as adding the absolute value of that quantity.

When an experiment is carried out under the conditions of temperature and pressure suggested by this contraction vertex, a yield of 82.68% is obtained. This is better than the response at the rejected vertex $\mathbf{W}$ (81.76%) but not as good as the other vertexes in the simplex (83.11 and 84.62%). But again, as far as the variable-

Simplex No. _5_ → _6_	Factor		Response	Rank	Vertex Number	Times Retained
	X_1	X_2	Response	Rank	Vertex Number	Times Retained
Coordinates of	56.93	71.07	84.62	B	9	2
retained vertexes	49.50	60.12	82.68	N	11	1
Σ	106.43	131.19				
$\bar{P} = Σ/k$	53.22	65.60				
W	37.60	65.90	83.11	W	6	3
$(\bar{P} - W)$	15.62	-0.30				
$R = \bar{P} + (\bar{P} - W)$	68.84	65.30	83.49	R	12	0
$(\bar{P} - W)/2$						
$C_W = \bar{P} - (\bar{P} - W)/2$				C_W		0
$C_r = \bar{P} + (\bar{P} - W)/2$				C_r		0
$E = R + (\bar{P} - W)$				E		0

Figure 4.12 Worksheet showing reflection only.

size simplex algorithm is concerned, it is irrelevant how good the response is at this C_W vertex. If a contraction is carried out, it is retained no matter what the response might be.

The vertex numbers listed on the worksheet (Figure 4.10) are "9" for **B**, "6" for **N**, and "5" for **W**. The reflection vertex **R** is vertex #10. The contraction vertex C_W becomes vertex #11. As before, whenever the coordinates of a vertex are calculated, that vertex receives the next sequential vertex number, whether it is retained in the next simplex or not.

The times retained column contains the value "1" for the best vertex **B**, "2" for the next-to-the-worst vertex **N**, and "3" for the wastebasket vertex **W**. The new vertexes **R** and C_W are each given a times retained value of "0."

The vertexes that will *not* appear on the next worksheet are **W** and **R**. The wastebasket vertex **W** never appears on the next worksheet (Rule 3). If a contraction is carried out, the reflection vertex **R** is never carried to the next worksheet.

The vertexes that *will* appear on the next worksheet are **B**, **N**, and C_W. The next-to-the-worst vertex **N** is always transferred into the wastebasket row labeled **W** on the next worksheet (Rule 3). Of the remaining two transferred vertexes (**B** and C_W in the current simplex), the current C_W vertex is *not* the better of the two. Thus, the **B** vertex will remain the **B** vertex on the next worksheet. The current C_W vertex will become the **N** vertex on the next worksheet.

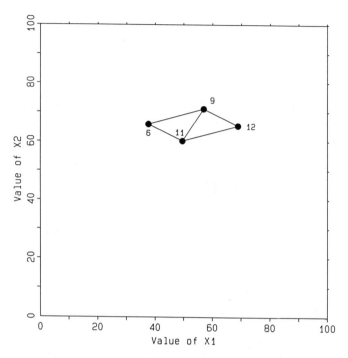

Figure 4.13   Graph of reflection only.

    Figure 4.11 summarizes the reflection followed by the C_W contraction. Vertex #10 (the reflection **R**) will not be retained and is connected to the original simplex with dashed lines. Vertex #11 (the contraction C_W) will be retained and is connected to the original simplex with solid lines to form the fifth complete simplex.

Reflection Only

Figure 4.12 shows a worksheet for carrying out the next simplex move, "Simplex No. *5 → 6*." Figure 4.13 is a graph of the resulting operations. Vertex #6 is now the wastebasket vertex and appears in the **W** row. Vertex #9 still gives the best response and appears in the top **B** row. Vertex #11 gives the next-to-the-worst response and appears in the **N** row. The reflection vertex **R** is located at a temperature of 68.84 and a pressure of 65.30.

    Notice that on this worksheet the worst vertex is not in the row labeled **W**. The next-to-the-worst vertex **N** has a response that is worse than the wastebasket response **W**. This happens occasionally with the simplex algorithms as they are presented in this text. This is why, after the first simplex, the row labeled **W** is usually called the "wastebasket vertex," not necessarily the "worst vertex." We will use the terms "worst" and "wastebasket" interchangeably to refer to the **W** vertex, even though its response might not always be the worst. (Similarly, the vertex **N** is not always the "next-to-the-worst vertex", but we will still call it by that name.)

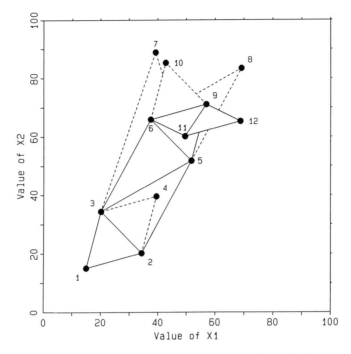

Figure 4.14 Summary of all simplex moves for the worked example.

When an experiment is carried out under the conditions of temperature and pressure suggested by this reflection vertex **R**, a yield of 83.49% is obtained. The response at **R** is not better than the response at **B**, so an expansion is not indicated. The response at **R** is not worse than the response at **N**, so a contraction is not indicated. This is a simple reflection move and is treated the same as a fixed-size simplex move.

The vertex numbers listed on the worksheet (Figure 4.12) are "9" for **B**, "11" for **N**, and "6" for **W**. The reflection vertex **R** is vertex #12.

The times retained column contains the value "2" for the best vertex **B**, "1" for the next-to-the-worst vertex **N**, and "3" for the wastebasket vertex **W**. The new vertex **R** is given a times retained value of "0."

The vertex that will *not* appear on the next worksheet is **W**. The wastebasket vertex **W** never appears on the next worksheet (Rule 3).

The vertexes that *will* appear on the next worksheet are **B**, **N**, and **R**. The next-to-the-worst vertex **N** is always transferred into the wastebasket row labeled **W** on the next worksheet (Rule 3). Of the remaining two transferred vertexes (**B** and **R** in the current simplex), the current **B** vertex will remain the **B** vertex on the next worksheet. The current **R** vertex will become the **N** vertex on the next worksheet.

Figure 4.13 summarizes the simple reflection move. Vertex #12 (the reflection **R**) will be retained and is connected to the original simplex with solid lines to form the sixth complete simplex.

Table 4.2 Self-Test for the Variable-Size Simplex Algorithm

Vertex	Factor x_1	Factor x_2	Response y_1
1. **I**	20.00	20.00	34.14
2. **I**	29.66	22.59	38.29
3. **I**	22.59	29.66	38.43
4. **R**	32.25	32.25	49.25
5. **E**	38.37	38.37	76.11
6. **R**	31.30	45.44	74.17
7. **R**	47.08	54.15	83.95
8. **E**	59.33	66.40	40.69
9. **R**	54.15	47.08	89.84
10. **E**	65.58	47.90	83.08
11. **R**	62.87	62.87	41.04
12. $\mathbf{C_W}$	44.49	44.49	88.59
13. **R**	51.57	37.42	90.42
14. **E**	53.81	29.06	83.14
15. **R**	61.23	40.01	94.86
16. **E**	69.59	37.77	93.85
17. **R**	67.00	28.11	94.87
18. **E**	73.43	18.63	88.16
19. **R**	85.03	28.46	91.63
20. $\mathbf{C_R}$	76.66	30.70	95.55
21. **R**	74.07	21.04	91.29
22. $\mathbf{C_W}$	70.71	33.59	96.34
23. **R**	80.37	36.18	86.59
24. $\mathbf{C_W}$	70.34	30.13	96.41
25. **R**	64.39	33.02	95.64
26. $\mathbf{C_R}$	67.46	32.44	97.30

Summary of the Variable-Size Simplex Moves

Figure 4.14 summarizes the variable-size simplex moves. The initial simplex consists of vertexes 1–2–3.

The simplex begins with a reflection followed by a successful expansion. The second simplex contains the vertexes 2–3–5.

The next move is a reflection followed by a failed expansion. The third simplex contains the vertexes 3–5–6.

The following move is a reflection that gives poor results, but not as poor as the result that was encountered at **W**. Thus, a $\mathbf{C_R}$ contraction results in the fourth simplex containing vertexes 5–6–9.

The move after that is a reflection that results in a $\mathbf{C_W}$ contraction. The response at **R** was worse than the result that was encountered on the **W** side. The fifth simplex consists of vertexes 6–9–11.

Finally, simple reflection results in the simplex 9–11–12.

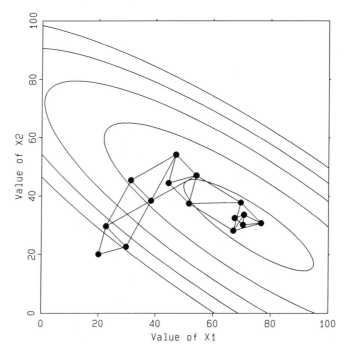

Figure 4.15 Summary of all simplex moves for the self-test for the variable-size simplex algorithm.

SELF-TEST FOR THE VARIABLE-SIZE SIMPLEX ALGORITHM

Make 12 copies of the two-factor variable-size simplex worksheet shown in Figure 4.3. Use these worksheets to verify the calculations of the variable-size sequential simplex optimization summarized in Table 4.2. Verify the graphical results shown in Figure 4.15. This optimization is discussed further in Chapter 5.

REFERENCES

1. J. A. Nelder and R. Mead, "A simplex method for function minimization," *Computer J.*, **7**, 308–313 (1965).
2. W. Spendley, G. R. Hext, and F. R. Himsworth, "Sequential application of simplex designs in optimisation and evolutionary operation," *Technometrics*, **4**(4), 441–461 (1962).

Chapter 5

Comments on Fixed-Size and Variable-Size Simplexes

SELF-TESTS FOR THE FIXED- AND VARIABLE-SIZE SIMPLEXES

p 94 *p 73*

In Chapter 3 you were asked to verify the calculations of the fixed-size sequential simplex optimization summarized in Table 3.2. Similarly, in Chapter 4 you were asked to verify the calculations of the variable-size sequential simplex optimization summarized in Table 4.2. Graphic summaries of these fixed- and variable-size optimizations are shown in Figures 5.1 and 5.2.

Note that each of the two optimizations used the same starting simplex with vertexes at (20.00, 20.00), (29.66, 22.59), and (22.59, 29.66), which gave responses of 34.14, 38.29, and 38.43, respectively. The two optimizations were carried out on the same response surface to allow comparison of the progress of the two different types of simplex under essentially identical conditions.

Comparing the Paths of the Two Simplexes

The fixed-size simplex (Figure 5.1) required approximately 22 vertexes to circle. Actually, circling was demonstrated when the coordinates of vertex #21 were calculated (see Table 3.2). Vertex #21 overlaps vertex #13 (at approximately the 11:00 position of the final clock-like hexagon in Figure 5.1) and confirms that the final simplex will circle in a clockwise manner about the optimum.

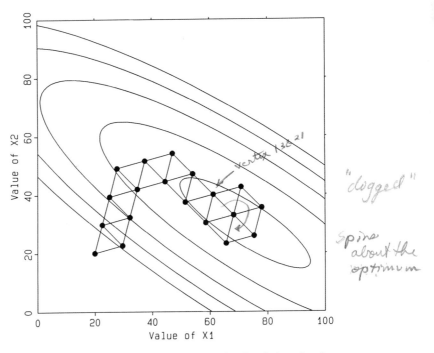

Figure 5.1 Graphic summary of self-test for the fixed-size simplex.

(see Table 3.2; p 73)

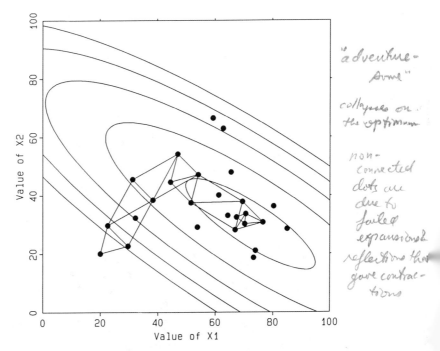

Figure 5.2 Graphic summary of self-test for the variable-size simplex.

(See Table 4.2; p94)

97

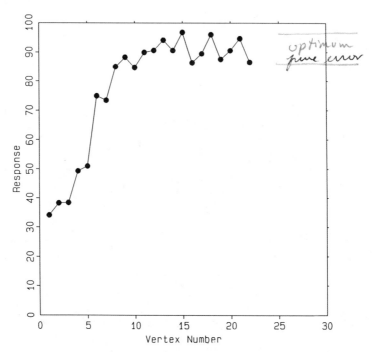

Figure 5.3 Response vs. vertex number for the fixed-size simplex self-test results.

The replicate experiments at vertexes #20, #21, and #22 were unnecessary but serve to show something about the repeatability of the experimental process. Notice that vertexes #20 and #14 at the 9:00 position have identical coordinates but gave slightly different responses: 90.55 and 90.59, respectively (Table 3.2). Similarly, vertexes #21 and #13 have identical coordinates but gave quite different responses: 94.67 and 94.12, respectively. Finally, vertexes #22 and #16 at the 1:00 position have identical coordinates but gave responses of 86.49 and 86.31, respectively. This lack of repeatability surprises some researchers, but it is a generally accepted statistical observation: when replicate experiments are carried out under supposedly identical conditions, different results are usually obtained. The name statisticians give to this variation is "pure error," or "purely experimental uncertainty."

The variable-size simplex (Figure 5.2) required about the same number of vertexes to bring it into the region of the optimum with the same precision as the fixed-size simplex. The variable-size simplex was allowed to run for 26 vertexes, but the last few vertexes contracted into a domain much tighter than the final hexagon of the fixed-size simplexes. None of the variable-size simplex vertexes overlaps, so it is not possible to estimate purely experimental uncertainty.

The two simplexes followed approximately the same path to the optimum. The path of the fixed-size simplex (Figure 5.1) was narrower and denser than the path of the variable-size simplex. The fixed-size simplex vertexes occur regularly over a ribbon-like pattern. In contrast, the variable-size simplex (Figure 5.2) tended to move rapidly over part of its path (this produced sparser information in

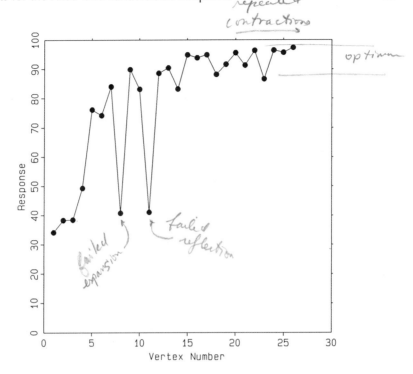

repeated contractions

optima

failed expansion *failed reflection*

Figure 5.4 Response vs. vertex number for the variable-size simplex self-test results.

those regions) and move slowly over other parts of its path (this produced denser information in those regions). The variable-size simplex also tended to scatter experiments about its path with failed expansions and with reflections that resulted in contractions.

The fixed-size simplex might be characterized as "dogged" and the variable-size simplex as "adventuresome."

Response vs. Vertex Number

Figure 5.3 plots response as a function of vertex number for the fixed-size simplex optimization. Figure 5.4 shows a similar plot for the variable-size simplex. In each case, the response started out relatively low at approximately 35, increased rapidly at first, then more slowly, and finally stabilized at approximately 95. This behavior is expected and is seen in Figures 5.1 and 5.2.

A relatively small simplex will quickly sense a direction of steep ascent and climb up onto a ridge; this explains the initial rapid increase in response in Figures 5.3 and 5.4. Once on the ridge, the simplex orients itself to the ridge and begins to move along the ridge toward better response; this accounts for the slower increase in response. Finally, as the simplex converges on the region of the optimum, it either spins about the optimum (fixed-size simplex) or collapses onto the optimum

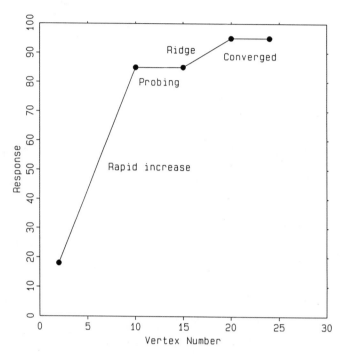

Figure 5.5 General behavior of small initial simplexes. Gains in response are initially rapid, but the rate of return decreases as the simplex probes to find the ridge and then moves along the shallower ridge to find the optimum.

(variable-size simplex); in either case, this is the reason for the final relatively constant response.

The pattern of the last seven responses at the right side of Figure 5.3 deserves comment. Note the pattern of three increasing responses followed by another set of three increasing responses and a final low response. If the simplex were allowed to continue, the last six points in Figure 5.3 would repeat (within purely experimental uncertainty). This is evident from Figure 5.1 where, in the final hexagonal pattern starting at the 1:00 position, responses of low, higher, very high, low, higher, and very high are observed as the clockwise-circling vertexes position themselves low on the side of the response surface, slightly higher up the side of the response surface, on the ridge, low on the side of the response surface, slightly higher up the side of the response surface, on the ridge, and so on.

The pattern of responses at the right side of Figure 5.4 also deserves comment. There is no obviously repeating pattern of responses here. Instead, the range of responses seems to be tightening (with one exceptional excursion toward low response at vertex #23). This tightening of responses is a consequence of the repeated contractions shown in Figure 5.2, which cluster the vertexes toward the optimum. Because the response surface is relatively flat in the region of the optimum, small variations in the factor levels produce small differences in response, and the responses get more and more similar as the vertexes get closer and closer together.

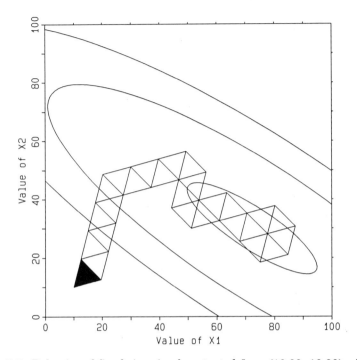

Figure 5.6 Behavior of fixed-size simplex started from (10.00, 10.00) with step sizes of +10.00 in each factor using the tilted algorithm for setting up the initial simplex.

The low responses at vertexes #8 and #11 in Figure 5.4 are the results of a failed expansion (vertex #8) and a failed reflection (vertex #11). Both of these vertexes represent excursions of the simplex over the ridge and down the other side (see Figure 5.2). The variable-size simplex uses the information from these experiments to begin to contract onto the ridge and direct itself along the ridge toward the optimum.

GENERAL BEHAVIOR OF SMALL STARTING SIMPLEXES

The previous discussion of response vs. vertex number allows some general observations on the behavior of relatively small simplexes started far from the optimum (such as those shown in Figures 5.1 and 5.2). Figure 5.5 shows this general behavior.

Initially, the simplex climbs the side of the response surface and produces very rapid increases in response. The side is the steepest part of the response surface, and the slope will be greatest there. A given amount of change in the factor levels here will produce the greatest change in response. This is indicated by the section of Figure 5.5 labeled "Rapid increase."

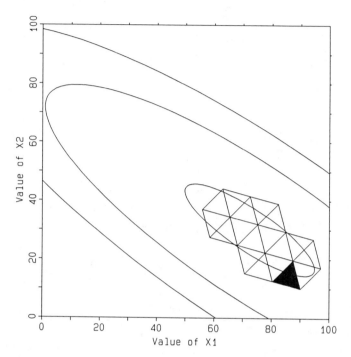

Figure 5.7 Behavior of fixed-size simplex started from (90.00, 10.00) with step sizes of -10.00 in factor x_1 and $+10.00$ in x_2.

After the simplex has climbed the side of the response surface, it finds itself on the ridge. The simplex must now investigate this region and gain enough information to determine the direction of the ridge so it can move along the ridge toward the optimum. This probing of the response surface to find the direction of the ridge does not produce very dramatic increases in response. This is indicated by the flat region of Figure 5.5 labeled "Probing."

When the simplex has found the direction of the ridge, it begins to move along the ridge toward the optimum. But the slope along the ridge is shallower than the slope up the side of the response surface. Thus, the gains in response come more slowly. This is indicated by the section of Figure 5.5 labeled "Ridge."

Finally, when the simplex has found the optimum, the responses become more or less constant, and no further improvements in response are obtained. This is indicated by the section of Figure 5.5 labeled "Converged."

It is important to realize that ridges usually have shallower slopes than the sides of response surfaces.

Some researchers use the sequential simplex for the first time and are delighted by the rapid increases in response it gives them initially. But later, as the simplex probes around to orient itself on a ridge, the researchers become disenchanted. After all, the simplex had been giving spectacular improvements in response, and now it is just sitting there giving about the same response. When it does begin to move again, the rate of improvement is considerably less than it was before.

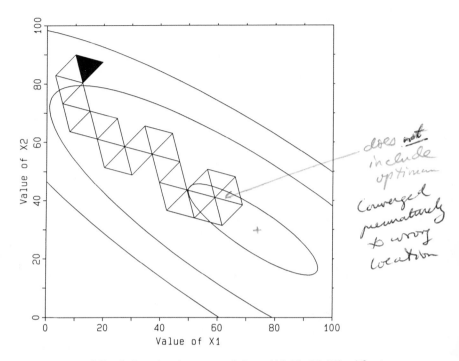

Figure 5.8 Behavior of fixed-size simplex started from (10.00, 90.00) with step sizes of +10.00 in factor x_1 and −10.00 in x_2.

Is the simplex broken? Has it converged? What is wrong? Why does it not behave as it did before? The answer should be clear now. The simplex is simply behaving as it is supposed to behave (see Figure 5.5). As it moves up the steep side of the response surface, the improvements in response are greater than the improvements found as it moves along the shallower ridge.

Behavior from Different Starting Locations

Figures 5.6–5.9 show the behavior of fixed-size simplexes starting from different corners of factor space on the same response surface as Figures 5.1 and 5.2.

In Figure 5.6, the fixed-size simplex is started from (10.00, 10.00) with step sizes of +10.00 in each factor using the tilted algorithm for setting up the initial simplex. (This algorithm is discussed in Chapter 6.) This initial simplex is shown as a filled triangle. As expected, the simplex rapidly climbs the side of the hill, turns onto the ridge, progresses toward the top of the ridge and finally overshoots, circles partially in a clockwise fashion to swing back onto the ridge, circles partially in a counter-clockwise manner to stay on the ridge, moves parallel with the ridge for a few moves, and then circles clockwise into convergence. The gains produced initially are very large as the simplex moves up the side of the hill; the gains come more slowly as the simplex moves along the ridge into the region of the optimum.

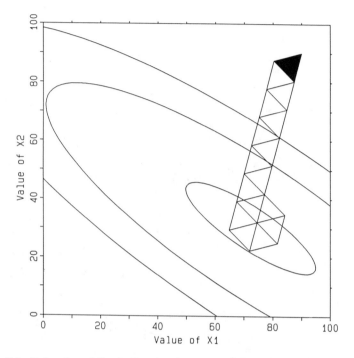

Figure 5.9 Behavior of fixed-size simplex started from (90.00, 90.00) with step sizes of −10.00 in each factor.

In Figure 5.7, the fixed-size simplex is started from (90.00, 10.00) with step sizes of −10.00 in factor x_1 and +10.00 in x_2. The simplex makes an initial reflection toward higher levels of x_1 and circles completely in a counterclockwise fashion. But the response at the seventh vertex is slightly better than the response at the center of the first hexagon, so the simplex begins to move in a clockwise manner to form a second hexagon (part of it overlaps with the first hexagon). This simplex soon breaks free and moves in a counterclockwise fashion again until it forms a third complete hexagon (part of it overlaps with the second hexagon). Finally, the simplex breaks free again and circles clockwise about the optimum region. In this example, the initial rapid gains in response are not seen: the initial simplex is already on the ridge close to the optimum, so only gradual increases in response are observed as the simplex slides over and straddles the optimum.

In Figure 5.8, the fixed-size simplex is started from (10.00, 90.00) with step sizes of +10.00 in factor x_1 and −10.00 in x_2. The simplex slides onto the ridge and moves along it until it circles in a region near the optimum. In this example, there are initially some rapid gains in response, but most of the remaining improvement is gradual.

Finally, in Figure 5.9, the fixed-size simplex is started from (90.00, 90.00) with step sizes of −10.00 in each factor. In this example the simplex happens to have a fortunate orientation. It is "pointed" directly toward the optimum, and rapid gains

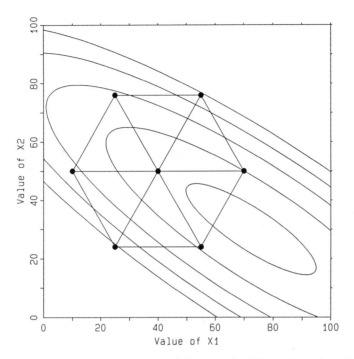

Figure 5.10 A fixed-size simplex stranded on a ridge. The orientation of the simplexes is such that the best (central) vertex lies on the ridge; all other vertexes lie lower on the sides of the response surface. All gradients point toward the central vertex. The information is not dense enough to detect the presence of the ridge.

in response are followed immediately by circling in the region of the optimum: there is no region of gradual improvement in the progress of this simplex.

Many persons see this example and think, "Gee. All I'd have to do is orient the initial simplex so it's pointed toward the optimum and it will be very efficient. It will get to the optimum almost instantly." Yes, but.... If the researcher knew where the optimum was, the researcher would not be using the simplex to try to get there. Remember once again, the contours of constant response are initially unknown to the researcher. They must be discovered by experiment. Behavior such as that shown in Figure 5.9 occurs only by chance.

BECOMING STRANDED ON A RIDGE

The optimum for the response surface used in this chapter is located at $x_1 = 72.33$, $x_2 = 30.07$. The fixed-size simplex shown in Figure 5.1 has done a good job of finding the region of the optimum: the final circling simplexes envelop the optimal coordinates. Similarly, the final simplexes shown in Figures 5.6, 5.7, and 5.9 surround the optimum. But the final simplexes in the example of Figure 5.8 do not

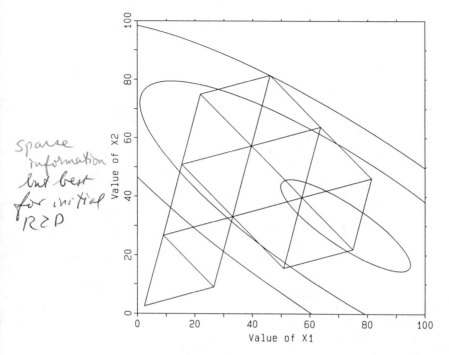

Handwritten note in left margin: Sparse information but best for initial RZP

Figure 5.11 Behavior of large initial fixed-size simplex started from (2.5, 2.5) with step sizes of +25.00 in each factor.

encompass the optimum. The simplex has converged into a region that does not include the optimum.

Figure 5.10 offers some insight as to why fixed-size simplexes can become stranded on ridges. By chance, the orientation of the simplexes in this figure is such that the best (central) vertex lies on the ridge and all other vertexes lie lower on the sides of the response surface. Thus, all gradients point toward the central vertex. The information from the vertexes is not dense enough to detect the presence of the ridge. This is clearly the case in Figure 5.8 as well, and the simplexes converge on the ridge some distance from the actual optimum.

Premature convergence of the fixed-size simplex can be a problem. Fortunately, the problem can be minimized by a modified fixed-size simplex strategy or by using the variable-size simplex instead.

STRATEGIES FOR USING
THE FIXED-SIZE SIMPLEX

The original fixed-size simplex was introduced as an alternative EVOP method for increasing the productivity of existing industrial processes. As such, it was intended that the simplexes be small, so as not to perturb the industrial process too much.

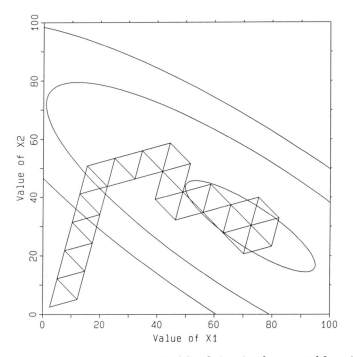

Figure 5.12 Behavior of moderate initial fixed-size simplex started from (2.5, 2.5) with step sizes of +10.00 in each factor.

Large Initial Simplex

If the initial simplex is relatively large, then the fixed-size simplex will make rapid progress across the response surface and will quickly locate the region of the optimum. This is illustrated in Figure 5.11.

Recall that the exact shape of the response surface is not known to the researcher initially. The response surface must be discovered by experiment. The elliptical contours of constant response shown in Figure 5.11 and subsequent figures simply serve to show us what the underlying response surface actually looks like and to help us understand why the simplex moves as it does. The simplexes (and ordinary researchers who use them) must "learn as they go."

In Figure 5.11, the initial vertex is located at $x_1 = 2.50$, $x_2 = 2.50$. The simplex quickly moves up onto the ridge and converges in a counterclockwise fashion around the region of the optimum. A total of 13 vertexes is shown.

The rapid progress of a large, fixed-size simplex across a response surface is clearly an advantage in a research or development environment where it is often necessary to quickly reach the region of the optimum.

Although this rapid progress also might seem to be attractive in a manufacturing or production environment, instead it is often a disadvantage. Practical safety considerations dictate that the simplex should not be allowed to be large – small

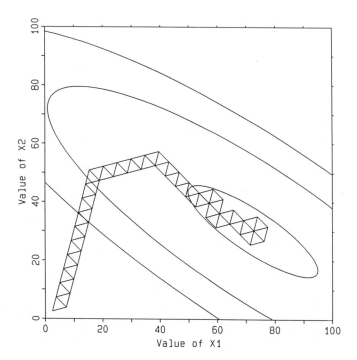

Figure 5.13 Behavior of small initial fixed-size simplex started from (2.5, 2.5) with step sizes of +5.00 in each factor.

simplexes reduce the risks of encountering unexpectedly hazardous experimental conditions.

One disadvantage that results from using a large, fixed-size simplex is that the location of the actual optimum is not known with very much precision. In Figure 5.11, we would suspect that the optimum *probably* lies somewhere within the final circling simplex (the final hexagon in Figure 5.11), but we would not know very precisely just where it is. (Remember again that the only experimental information we would have at this point is the information from the 13 simplex vertexes shown in Figure 5.11. We would not yet have the omnipotent view given by the elliptical contours of constant response.)

There is, however, one additional advantage of this type of experimentation. Even though we do not know very precisely just where the optimum is, and even though we have not done very many experiments, the experiments have been collected over a rather broad region of factor space. If we were to use this set of data to fit an empirical mathematical model, something like a full second-order polynomial model, e.g.,

$$y_{1i} = \beta_0 + \beta_1 x_{1i} + \beta_2 x_{2i} + \beta_{11} x_{1i}^2 + \beta_{22} x_{2i}^2 + \beta_{12} x_{1i} x_{2i} + r_{1i}$$

we would probably find that the fitted model does an adequate job of describing most of the response surface.

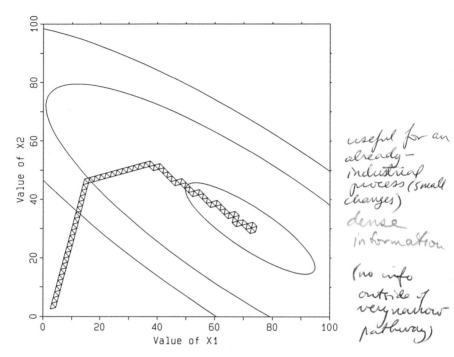

useful for an already-industrial process (small changes)

dense information

(no info outside of very narrow pathway)

Figure 5.14 Behavior of tiny initial fixed-size simplex started from (2.5, 2.5) with step sizes of +2.00 in each factor. *May have to repeat vertexes if responses at any given simplex are too similar*

Moderate Initial Simplex

In Figure 5.12, it is clear that using a moderate initial simplex will give better precision in locating the optimum. (The initial vertex is again located at $x_1 = 2.50$, $x_2 = 2.50$.) The optimum probably lies somewhere within the final circling simplex (the last, clockwise hexagon at the right). Even though we do not know exactly where the optimum is located, we do know that it probably lies within a smaller area of uncertainty that we obtained in Figure 5.11. But there is a price to be paid for this improved precision. That price is measured by the number of extra vertexes that were required to move the smaller simplex into the region of the optimum.

Small Initial Simplex

If we use a small initial simplex, the optimum can be located even more precisely (see Figure 5.13). A still smaller simplex locates the optimum with even better precision (see Figure 5.14), but the price now becomes severe. Many more experiments are required to reach the region of the optimum. As shown in Figure 5.14, the optimum has been located very precisely, but a very large number of experiments were required.

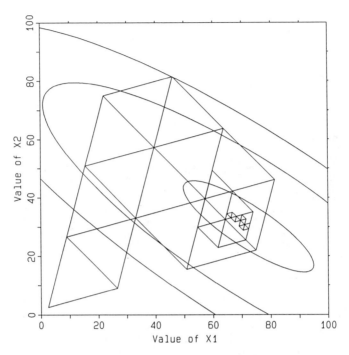

Figure 5.15 Behavior of sequential sequential fixed-size simplexes. Large initial fixed-size simplex started from (2.5, 2.5) with step sizes of +25.00 in each factor. Subsequent moderate fixed-size simplex started from previous best with step sizes of +10.00 in each factor. Final tiny fixed-size simplex started from previous best with step sizes of +2.00 in each factor.

There is another disadvantage in using a very small simplex. If the data from Figure 5.14 were used to fit an empirical model, the model would probably give excellent prediction of the response surface over the path that has been traveled, but prediction in unexplored regions would be subject to very large uncertainties. Contrast this with the results of using a very large simplex shown in Figure 5.11. In Figure 5.14 there is very dense information over the path traveled, but there is no information in other regions. In Figure 5.11 there is sparse information over the whole response surface.

 For a process still in research and development, the very large simplex of Figure 5.11 might be preferred. For an industrial process that is already in production, the very small simplex of Figure 5.14 might be preferred. If it is necessary to optimize an existing industrial process, then each move must be very conservative. It might be necessary to run the vertexes at each simplex several times before it is possible to make a clear decision as to which vertex is actually the worst, but in such a situation speed is not usually important. After all, if the process is running it is already making money and you can afford to move slowly and cautiously.

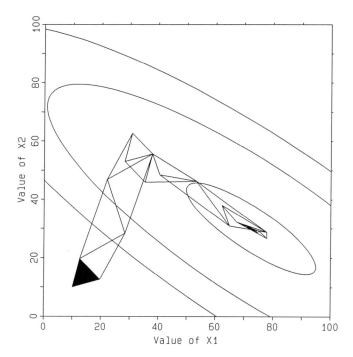

Figure 5.16 Behavior of small variable-size simplex started from (10.00, 10.00) with step sizes of +10.00 in each factor using the tilted algorithm for setting up the initial simplex.

Sequential Sequential Simplexes

Long [1] suggested an ingenious way of using sequences of sequential fixed-size simplexes. In Figure 5.15, a large initial simplex has been used to quickly get into the region of the optimum. A smaller simplex was then started with the previous best vertex as its starting point and allowed to converge to a more precise optimum. A still smaller simplex was then started at that new best vertex. This repetitive application of smaller and smaller simplexes is one way to make use of the best features of large and small simplexes. *(contractions)*

STRATEGIES FOR USING THE VARIABLE-SIZE SIMPLEX

The contractions suggested in Figure 5.15 are carried out more naturally and automatically with the variable-size simplex of Nelder and Mead [2]. In this section we compare and contrast the behavior of the variable-size simplex algorithm with the fixed-size algorithm.

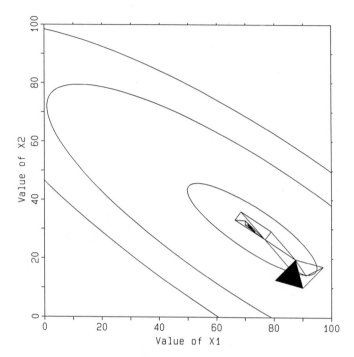

Figure 5.17 Behavior of small variable-size simplex started from (90.00, 10.00) with step sizes of -10.00 in factor x_1 and $+10.00$ in x_2.

Behavior from Different Starting Locations

Figures 5.16–5.19 show the behavior of variable-size simplexes starting from different corners of factor space on the same response surface as shown in Figures 5.6–5.9.

In Figure 5.16, the variable-size simplex is started from (10.00, 10.00) with step sizes of $+10.00$ in each factor using the tilted algorithm for setting up the initial simplex. This initial simplex is shown as a filled triangle. As expected, the simplex rapidly expands up the side of the hill, contracts onto the ridge, accelerates along the ridge, and finally collapses about the optimum. As expected from the discussion of Figure 5.4, the gains produced initially are very large as the simplex moves up the side of the hill; the gains come more slowly as the simplex moves along the ridge and onto the optimum.

In Figure 5.17, the variable-size simplex is started from (90.00, 10.00) with step sizes of -10.00 in factor x_1 and $+10.00$ in x_2. The simplex makes an initial simple reflection toward higher levels of x_1. The next two moves are contractions that orient the simplex toward the direction of the ridge. The simplex accelerates and directs itself toward the optimum and collapses there. In this example (as in Figure 5.7), the initial rapid gains in response are not seen: the initial simplex is already

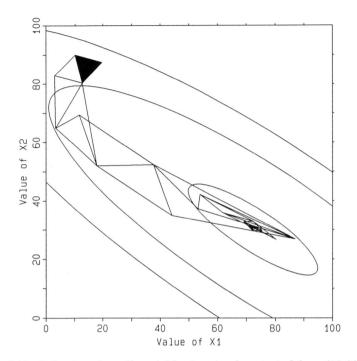

Figure 5.18 Behavior of small variable-size simplex started from (10.00, 90.00) with step sizes of +10.00 in factor x_1 and −10.00 in x_2.

on the ridge close to the optimum, so only gradual increases in response are observed as the simplex slides over and collapses onto the optimum.

In Figure 5.18, the variable-size simplex is started from (10.00, 90.00) with step sizes of +10.00 in Factor x_1 and −10.00 in x_2. The simplex swings onto the ridge and slides smoothly into the region of the optimum where it collapses. In this example, there are initially some rapid gains in response, but most of the remaining improvement is gradual. Note that in Figure 5.8, the fixed-size simplex converged prematurely; this is not the case with the variable-size simplex.

Finally, in Figure 5.19, the variable-size simplex is started from (90.00, 90.00) with step sizes of −10.00 in each factor. In this example, the simplex is "aimed" almost directly at the optimum and rapid gains in response are followed almost immediately by collapsing into the region of the optimum: there is no region of gradual improvement in the progress of this simplex.

Difficulties with Small Initial Variable-Size Simplexes

In these four examples (Figures 5.16–5.19), the simplex follows a relatively narrow path toward the optimum. If the response surface were to be modeled from the

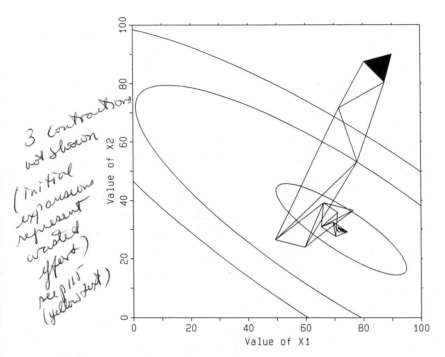

(handwritten marginal note) 3 contractions not shown (Initial expansions represent wasted effort) see p 115 (yellow text)

Figure 5.19 Behavior of small variable-size simplex started from (90.00, 90.00) with step sizes of −10.00 in each factor.

simplex data, then the path followed would be known with high confidence, but those regions not explored by the simplex would not be known very well. Predictions of response in these unexplored regions would be accompanied by large uncertainties.

There is another, subtler problem with the variable-size simplex optimizations shown in Figures 5.16–5.19. It is perhaps most evident in Figure 5.19. Note that the initial reflection is followed by a successful expansion that elongates the simplex and allows it to move more rapidly up the side of the response surface. This is followed by a second successful expansion that extends the simplex even farther in this favorable direction. A third successful expansion brings the expansion vertex just slightly over the hill. At this point the simplex is eight times as large as it was initially and is extended greatly in the direction that has been favorable up to this point. This has been achieved at the expense of three extra vertex evaluations that are not shown in Figure 5.19 (the three reflection vertexes).

The problem now is that the currently large simplex must get small again so it can find the ridge and move on toward the optimum. As seen in Figure 5.19, a C_R contraction is followed by a C_W contraction, which is then followed by a simple reflection. This simple reflection is followed by one more C_W contraction at which point the simplex now contains the same two-dimensional volume as the original simplex. These contractions back to the simplex's original size have been achieved

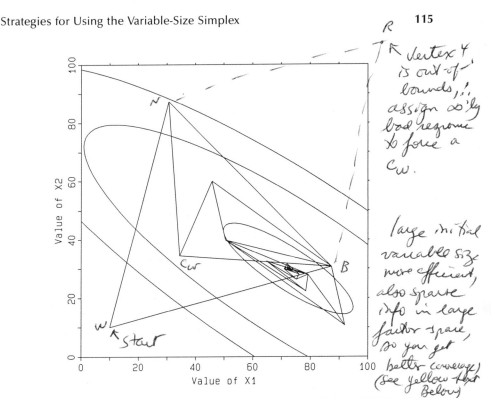

(handwritten margin note:) ↑ vertex 4 is out-of-bounds, assign so'ly bad regime to force a c_w.

(handwritten margin note:) large initial variable size more efficient, also sparse info in large factor space, so you get better coverage) (see yellow text below)

Figure 5.20 Behavior of large variable-size simplex started from (10.00, 10.00) with step sizes of +80.00 in each factor using the tilted algorithm for setting up the initial simplex.

at the expense of three rejected reflection vertex evaluations that are not shown in Figure 5.19.

It should be clear now that the supposed advantage of initial expansions to accelerate in a good direction has had to be followed by the disadvantage of later contractions to converge toward the optimum. In a sense, the initial expansions represent wasted effort.

Large Initial Variable-Size Simplexes

A few moments thought will suggest the following strategy: start with a large initial simplex and simply let it collapse onto the optimum. The idea is similar to Long's philosophy of sequential sequential fixed-size simplexes shown in Figure 5.15, where a large initial simplex optimization is followed by a smaller simplex optimization, which in turn is followed by a still smaller simplex optimization, and so on.

Starting with a large initial variable-size simplex is generally more efficient because it avoids the wasted effort of letting the simplex get large on its own. A large initial variable-size simplex has an additional benefit: sparse information

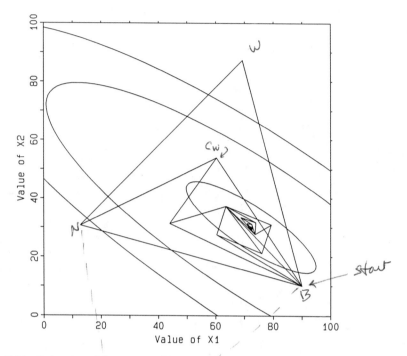

Figure 5.21 Behavior of large variable-size simplex started from (90.00, 10.00) with step sizes of -80.00 in factor x_1 and $+80.00$ in x_2.

will be obtained from a large region of factor space – models will be more robust and predictions made from them will be more certain.

Behavior from Different Starting Orientations

Figures 5.20–5.23 show the progress of large variable-size simplexes. In each example, the starting vertex is located at a different position in factor space and the step sizes are adjusted accordingly.

In Figure 5.20, the variable-size simplex is started from (10.00, 10.00) with step sizes of $+80.00$ in each factor using the tilted algorithm for setting up the initial simplex. The first move is a rejection of the vertex at (10.00, 10.00), which causes the reflection vertex to be out of bounds near the upper right side of the figure. If this out-of-bounds vertex is assigned an infinitely bad response (a reasonable decision), then the simplex algorithm will automatically carry out a C_W contraction that gives a vertex that is within bounds. Other contractions and reflections move the vertexes toward the optimum.

Figure 5.21 shows similar behavior for a variable-size simplex started from (90.00, 10.00) with step sizes of -80.00 in factor x_1 and $+80.00$ in x_2. Collapse onto the optimum is straightforward.

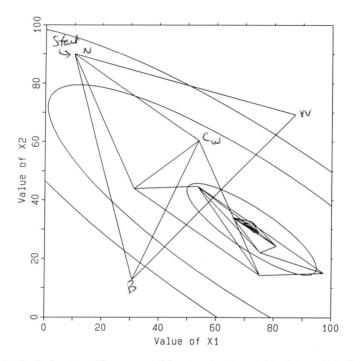

Figure 5.22 Behavior of large variable-size simplex started from (10.00, 90.00) with step sizes of +80.00 in factor x_1 and −80.00 in x_2.

In Figure 5.22, the variable-size simplex is started from (10.00, 90.00) with step sizes of +80.00 in factor x_1 and −80.00 in x_2. Unlike the initial simplexes in the previous two figures, this initial simplex does not contain the region of the optimum. Thus, after two contractions, a reflection, another contraction, and another reflection, the simplex finally envelops the optimum. Further moves collapse the simplex onto the optimum.

Finally, in Figure 5.23, the variable-size simplex is started from (90.00, 90.00) with step sizes of −80.00 in each factor. Collapse onto the optimum is direct.

In our experience, we have found the large initial variable-size simplex to be preferable for early research and development work, especially in projects in which there is very little prior knowledge. The simplex rapidly focuses the experimental effort into the region of the optimum.

General Behavior of Large Initial Variable-Size Simplexes

Figure 5.24 shows the possible behavior of response vs. vertex number for large initial variable-size simplexes. In general, it is similar to the behavior shown in Figure 5.5 for small initial simplexes.

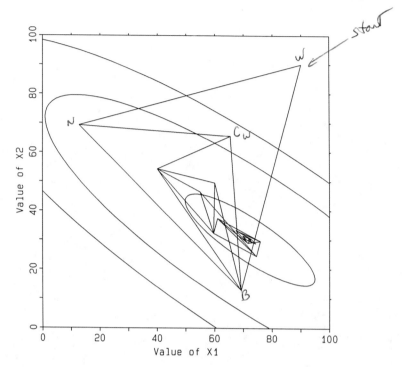

Figure 5.23 Behavior of large variable-size simplex started from (90.00, 90.00) with step sizes of −80.00 in each factor.

However, with a large initial simplex, it is likely that the responses from the vertexes will cover a fairly broad range. The response from one of the vertexes might even be close to the optimum. (This is seen, for example, in Figure 5.20.) If this occurs, then the simplex will spend its initial moves getting rid of the less desirable vertexes and replacing them with contraction vertexes that lie closer and closer to the best vertex and therefore have better and better response. This will produce an ever-narrowing band of good responses as the simplex progresses.

It is this behavior that is indicated by the shaded region in Figure 5.24. The shading is meant to convey the idea that the spread of response from the initial simplex might be very broad—from very poor response to (in this case) optimal response. As the simplex progresses, the range of responses narrows toward the better responses.

A CURIOUS SUCCESS STORY

At one of the short courses where we presented some of this material, a participant and his immediate supervisor came up to us during a break to talk about some of the things we had said. This participant strongly disagreed with us. He said something like this:

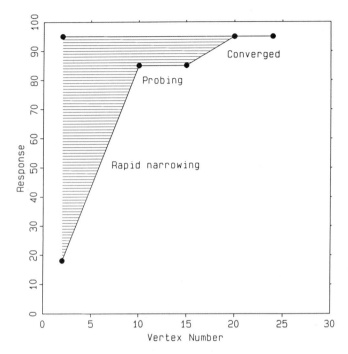

Figure 5.24 General behavior of large initial variable-size simplexes. Same as Figure 5.5 but with a new possibility indicated by the shaded region: one or more of the initial vertexes might be very good, in which case the remaining vertexes are pulled up the side of the hill toward the best vertex until the simplex converges.

"If I understand you correctly, you're suggesting that I use a large initial simplex that covers as much of the factor space as it can." We said, yes, that is what we were suggesting. He went on to say, "In my work I carry out organic synthetic reactions, and in one of these reactions I have to do an acid hydrolysis. I get about forty percent yield. I think pH might be an important factor for getting improved yields, but I do not think it makes any sense to vary the pH from zero to fourteen."

Before we could explain to him that an appropriately broad factor domain for acid hydrolysis might involve pH values from one to three, his supervisor broke in and said, "These guys told us to vary the pH all the way from zero to fourteen. Do what they say."

Well, now we were in a real mess. We did not want to contradict this supervisor, so we just swallowed hard, shrugged our shoulders, offered both of them a cup of coffee and a cookie, and mentioned the very cold weather they had been having in their city recently. The participant did not ask any more questions or make any more comments during the remainder of the course, and we did not have a chance to get back with him on the subject of pH domains.

We found out later that this participant was very angry with us, but he did what his supervisor said to do. He constructed a large initial simplex similar to the

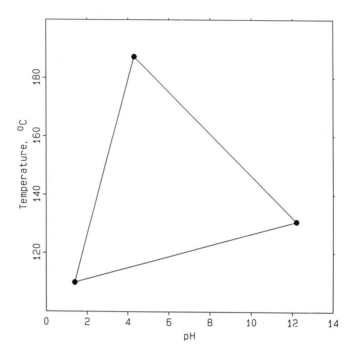

Figure 5.25 A large initial variable-size simplex in the factor space of pH and temperature.

one in Figure 5.25, with most of the vertexes at low pH, but with one of the vertexes clearly at a very high pH (basic, not acidic, conditions).

Now, if you were upset about being told to carry out acid hydrolyses at the factor combinations shown in Figure 5.25, and if you wanted to prove to your supervisor that this simplex was a bunch of nonsense, which experiment would you do first? You are right. You would do the experiment at high pH, the most nonacidic conditions. Well, this researcher carried out that experiment, and got 99% yield. Needless to say, he did not carry out experiments at any more factor combinations.

What happened here? In effect, the large initial simplex forced the researcher to consider factor combinations that he would never have considered otherwise. As it turns out in this case, a base hydrolysis worked much better than an acid hydrolysis. But he would never have tried base hydrolysis as long as he was convinced that acid hydrolysis was the method of choice.

There is a fundamental philosophical issue involved here. Do we do things a certain way because other researchers have found out that other ways do not work? Or do we do things a certain way only because other researchers have found that this way works pretty well and have not bothered to explore other possibilities?

The serendipitous effect of a large, sparse initial design has been noted recently by Hahn et al. [3], who state:

> a fractional factorial plan [similar in many ways to a simplex design] can be
> an effective mechanism for scanning the response surface to identify those

> regions that seem most promising for further study. . . . In fact, we have found this general scanning of the response surface, without necessarily fitting a model, to be one of the major advantages of the fractional factorial plans–and one often not sufficiently emphasized by texts.

The large initial simplex will not always result in such a nice success story, but if this happened only 5% of the time, that would open up some very interesting new ways of doing things.

REFERENCES

1. D. E. Long, "Simplex optimization of the response from chemical systems," *Anal. Chim. Acta*, **46**(2), 193–206 (1969).
2. J. A. Nelder and R. Mead, "A simplex method for function minimization," *Comput. J.*, **7**, 308–313 (1965).
3. G. J. Hahn, J. L. Bemesderfer, and D. M. Olsson, "Explaining experimental design fundamentals to engineers: A modern approach," Chapter 4 in C. L. Mallows, *Design, Data, and Analysis: by Some Friends of Cuthbert Daniel*, Wiley, New York, 1987, pp. 41–70.

Chapter 6

General Considerations

THE INITIAL SIMPLEX

The performance of the sequential simplex method depends to some extent on the size, orientation, and location of the initial simplex. Although experiments for the vertexes of the initial simplex can be supplied by the user, it is recommended that one of the formal setup algorithms discussed below (tilted or corner) be used to determine the experimental conditions for the initial simplex vertexes.

User-Supplied Initial Simplex

Occasionally, the user will want to supply the vertexes of the initial simplex. This might be the case if the user has theoretical information that suggests at least $k + 1$ potentially "good" starting conditions. These experimental conditions might be used to define the vertexes of the initial simplex. At other times, prior experiments might already have been carried out on the system and $k + 1$ of these experiments can be used as vertexes for the initial simplex. In this second case, the advantage is obvious: there is no need to carry out $k + 1$ new experiments—an "initial simplex" already exists.

Figure 6.1 is an example of a user-supplied simplex. One vertex is defined by an experiment carried out at ($x_1 = 60.00$, $x_2 = 20.00$). A second vertex is defined by the experiment carried out at ($x_1 = 30.00$, $x_2 = 40.00$). The third vertex is defined by the experiment at (65.00, 80.00).

Table 6.1 contains a tabular representation of the three vertexes (60.00, 20.00), (30.00, 40.00), and (65.00, 80.00) in the initial simplex shown in Figure 6.1. In this tabular representation there is one column for each of the k factors, and one row for each of the $k + 1$ vertexes. Table 6.1 is similar to the arrangement of vertexes in a simplex worksheet.

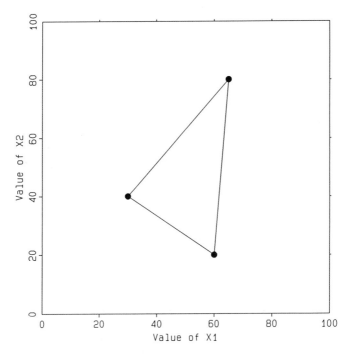

Figure 6.1 An example of a user-supplied initial simplex.

Figure 6.2 is another example of a user-supplied simplex. One vertex is defined by an experiment carried out at $(x_1 = 10.00, x_2 = 90.00)$. A second vertex is defined by the experiment carried out at $(x_1 = 50.00, x_2 = 50.00)$. The third vertex is defined by the experiment at $(40.00, 60.00)$. Table 6.2 contains the tabular representation of the three vertexes shown in Figure 6.2.

In Figure 6.2, the simplex does not span the two-dimensional factor space. Instead, the normally triangular two-dimensional simplex has collapsed into a one-dimensional line segment. The usual simplex calculations will cause the simplex to move, but the simplex is constrained to move in one dimension only – it cannot enter the full two-dimensional plane of the x_1–x_2 factor space. Normally, this is an undesirable situation. (However, see Mixture Design Experiments later in this chapter.)

Table 6.1 Tabular Representation of the Three Vertexes Shown in Figure 6.1

Vertex	Factor x_1	Factor x_2
1	60.00	20.00
2	30.00	40.00
3	65.00	80.00

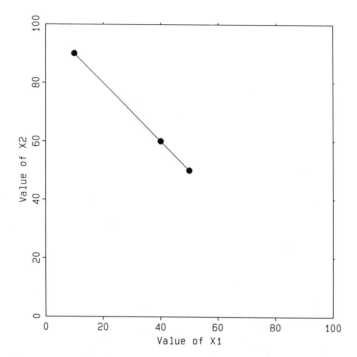

Figure 6.2 Another example of a user-supplied initial simplex.

   A simplex that exists entirely in a lower dimensional space is called a "degenerate" simplex. The vertexes of any degenerate simplex lie in a space that is of lower dimension than the factor space. Thus, the vertexes of a degenerate simplex are "cohypoplanar." In two or three dimensions, cohypoplanarity is easy to recognize (see, for example, Figure 6.2). In higher dimensions where geometry is harder to visualize, it is not usually possible to detect cohypoplanarity by inspection. Instead, mathematical tools are necessary.

   Cohypoplanarity arises because the vertexes are not linearly independent. In Figure 6.2 and Table 6.2, for example, any one of the vertexes can be obtained from the other two by simple linear algebraic operations. To illustrate, subtracting the coordinates of vertex #2 from the coordinates of vertex #1 gives $(-40.00, 40.00)$. Dividing this difference by 4 gives $(-10.00, 10.00)$. When this quotient is added to vertex #2, the result is $(40.00, 60.00)$, the coordinates of vertex #3.

Table 6.2 Tabular Representation of the Three Vertexes Shown in Figure 6.2

Vertex	Factor x_1	Factor x_2
1	10.00	90.00
2	50.00	50.00
3	40.00	60.00

The following mathematical operations can be used to determine if a set of vertexes is cohypoplanar:

1. Subtract the coordinates of one vertex from the coordinates of all other vertexes. Geometrically, this has the effect of translating the coordinate axes to the chosen vertex; alternatively, it has the effect of translating the simplex to the origin of the coordinate axes (similar to the calculation of $\bar{P}-W$ in Chapter 3).
2. Take the determinant of the square matrix formed by the remaining vertexes.
3. If this determinant is zero, then the simplex is cohypoplanar.

To illustrate, the coordinates of the vertexes in Figure 6.2 are

Vertex	Factor x_1	Factor x_2
1	10.00	90.00
2	50.00	50.00
3	40.00	60.00

Subtracting vertex #3 from the others gives

Vertex	Factor x_1	Factor x_2
1'	−30.00	30.00
2'	10.00	−10.00

The determinant of this 2×2 matrix is calculated by multiplying the elements of the main diagonal (running from upper left to lower right) and subtracting from it the product of the off-diagonal elements:

$$(-30)\times(-10) - (10)\times(30) = 300 - 300 = 0$$

A similar calculation can be carried out on the vertexes shown in Figure 6.1 for a noncohypoplanar simplex:

Vertex	Factor x_1	Factor x_2
1	60.00	20.00
2	30.00	40.00
3	65.00	80.00

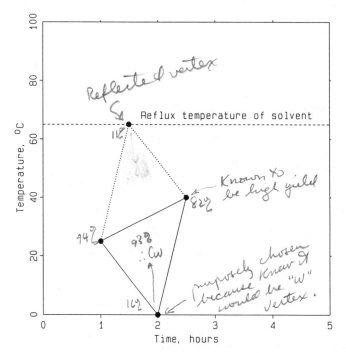

Figure 6.3 A user-supplied initial simplex that was set up in a purposeful manner.

Subtracting vertex #3 from the others gives

Vertex	Factor x_1	Factor x_2
1'	−5.00	−60.00
2'	−35.00	−40.00

The determinant of this 2×2 matrix is

$$(-5) \times (-40) - (-35) \times (-60) = 200 - 2100 = -1900$$

which is clearly not zero. Thus, the simplex exists in the full two-dimensional factor space and is not degenerate.

It is a little-known statistical fact that two-level screening designs such as the saturated fractional factorial [1, 2], Plackett-Burman [3], Hadamard [4–6], and Taguchi [7–9] are simplex designs. Thus, an initial screening design in q factors can often function as a user-supplied initial simplex for a sequential simplex optimization. (Conversely, an initial simplex in q factors can often be used as a screening design, although it might not have such statistically desirable properties as orthogonality and minimum variance estimates.) If only a subset k of the original q factors is used for subsequent optimization, care must be exercised that the selection of $k + 1$ factor combinations from the original $q + 1$ factor combinations does not give a degenerate simplex.

Purposeful Initial Simplex

Figure 6.3 shows an initial simplex that was set up in a purposeful or deliberate manner [10]. The application involved maximizing the percentage yield of desired product in an organometallic chemical reaction. Reaction time in hours (horizontal axis) and reaction temperature in Celsius degrees (vertical axis) were chosen as the two factors for a sequential simplex optimization. Preliminary experimentation had found relatively high yield (approximately 80%) at 2.5 hours reaction time and 40 °C reaction temperature, so this was used as one of the initial simplex vertexes.

The reaction was run in tetrahydrofuran, which has a boiling point of approximately 65 °C (and could not be exceeded). The lowest temperature possible with the experimental apparatus was ice temperature, 0 °C. Thus, although time had only a lower bound (0 hours), temperature had both a lower bound (0 °C) and an upper bound (65 °C). The investigators were interested in determining the yield at both the lower and the upper temperature boundary. They did not believe the yield would be very good at the lower temperature boundary, but expected it to be quite good at the higher temperature boundary. The investigators chose 2.00 hours and 0 °C as the conditions for the second vertex, and then carefully chose the coordinates of the remaining vertex: 1.00 hour and 25 °C.

By using this carefully crafted initial simplex, if their supposition was correct that the worst yield would occur at 0 °C, then rejecting that vertex would result in a reflection to 1.5 hours and 65 °C, the upper temperature boundary (shown by the dotted lines in Figure 6.3).

Table 6.3 General Tabular Representation of the $k + 1$ Vertexes in a k-Dimensional Simplex

Vertex	Factor x_1	Factor x_2	Factor x_3	Factor ...	Factor X_k
1	$v_{1,1}$	$v_{2,1}$	$v_{3,1}$	$\cdot$	$v_{k,1}$
2	$v_{1,2}$	$v_{2,2}$	$v_{3,2}$	$\cdot$	$v_{k,2}$
3	$v_{1,3}$	$v_{2,3}$	$v_{3,3}$	$\cdot$	$v_{k,3}$
4	$v_{1,4}$	$v_{2,4}$	$v_{3,4}$	$\cdot$	$v_{k,4}$
$\cdot$	$\cdot$	$\cdot$	$\cdot$	$\cdot$	$\cdot$
$\cdot$	$\cdot$	$\cdot$	$\cdot$	$\cdot$	$\cdot$
$\cdot$	$\cdot$	$\cdot$	$\cdot$	$\cdot$	$\cdot$
$k + 1$	$v_{1,k+1}$	$v_{2,k+1}$	$v_{3,k+1}$	$\cdot$	$v_{k,k+1}$

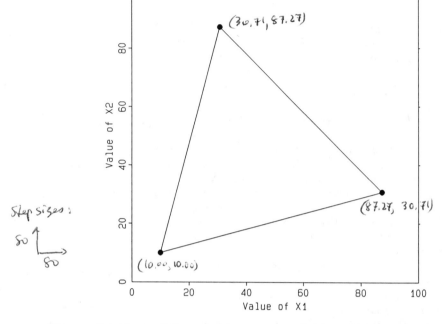

Figure 6.4 A two-factor simplex generated by the tilted initial simplex algorithm.

When the three initial vertexes were evaluated, the vertex representing previously used conditions gave a yield of 82%, the vertex at $0°C$ gave 16%, and the third vertex gave an intermediate yield of 74%. The investigators' supposition was correct, and the vertex at $0°C$ was rejected, resulting in a reflection to the boiling point of tetrahydrofuran. Much to the researchers' surprise, however, the yield at this temperature was only 11%! Thus, a C_W contraction was indicated. (After several more moves, a yield of approximately 93% was achieved and the simplex was halted.)

Tilted Initial Simplex

Table 6.3 shows a general tabular representation of the $k + 1$ vertexes in a k-dimensional simplex. There is one column for each of the k factors, and one row for each of the $k + 1$ vertexes. In Table 6.3, the symbol v stands for the value of a factor. The subscripts on v indicate, first, the factor (column) and, second, the experiment (row). This is opposite to the usual mathematical convention for subscripts on matrix elements, which refers to rows first, then columns; it is, however, consistent with—and Table 6.3 is analogous to—the experimental design matrix D in the statistical design of experiments literature [11]. The symbol $v_{2.3}$, for example, represents the value of the second factor in the third experiment.

Table 6.3 is similar to the arrangement of vertexes in a simplex worksheet. If Table 6.3 contains the vertex coordinates of the initial simplex, it is important that the determinant of its square submatrix be nonzero. One way to guarantee a nonzero determinant is to use a formal algorithm to construct the initial simplex.

The tilted initial simplex algorithm was first presented by Spendley, Hext, and Himsworth [12] and has been discussed in detail by Beveridge and Schechter [13]. Figure 6.4 shows a two-factor initial simplex generated by this algorithm.

For each factor dimension i, two quantities are specified: the coordinate of the "starting vertex" for that factor, S_i, and the "step size" for that factor, s_i. The coordinate of the "starting vertex" acts as a reference point for the generation of the other vertexes. It is usually chosen to be low in all factors, or might represent a "best guess" of where the optimum might be.

The "step size" is a measure of the span of the initial simplex in each factor dimension. If the starting vertex was chosen to be low in all factors, then the step size in each factor might be large to span each factor domain. If the starting vertex was chosen to represent a best guess of where the optimum might be, then the step size might be smaller to confine the initial simplex to the suspected region of optimum response. Step sizes can be negative.

[Some writers have suggested that the step size in each factor should be chosen so that the difference in response between the starting vertex and each of the other vertexes will be about the same. In this way, the factors can be normalized so that a unit (i.e., step size) change in each factor will produce similar changes in response. Unfortunately, this suggestion presumes a high level of prior knowledge about the system. If investigators were able to make use of this suggestion, they would already know enough about the system so that they would not have to carry out the initial simplex.]

Table 6.4 Tabular Representation of the $k + 1$ Vertexes in a Tilted Initial Simplex[a]

Vertex	Factor x_1	Factor x_2	Factor x_3	⋯	Factor X_k
1	S_1	S_2	S_3		S_k
2	S_1+p_1	S_2+q_2	S_3+q_3	⋅	S_k+q_k
3	S_1+q_1	S_2+p_2	S_3+q_3	⋅	S_k+q_k
4	S_1+q_1	S_2+q_2	S_3+p_3	⋅	S_k+q_k
·	·	·	·		·
·	·	·	·		·
·	·	·	·		·
$k + 1$	S_1+q_1	S_2+q_2	S_3+q_3		S_k+p_k

[a]S_i = starting coordinate for factor i.

s_i = step size for factor i.

$p_i = s_i [\sqrt{k+1} + k-1]/[k\sqrt{2}]$

$q_i = s_i [\sqrt{k+1} - 1]/[k\sqrt{2}]$

Table 6.5 Tabular Representation of the $k + 1$ Vertexes in a Corner Initial Simplex

Vertex	Factor x_1	Factor x_2	Factor x_3	·	Factor X_k
1	S_1	S_2	S_3	·	S_k
2	$S_1 + s_1$	S_2	S_3	·	S_k
3	S_1	$S_2 + s_2$	S_3	·	S_k
4	S_1	S_2	$S_3 + s_3$	·	S_k
·	·	·	·	·	·
·	·	·	·	·	·
·	·	·	·	·	·
$k + 1$	S_1	S_2	S_3	·	$S_k + s_k$

In the tilted initial simplex algorithm, two quantities are derived from the step size for each factor [12, 13]:

$$p_i = s_i[\sqrt{k+1} + k - 1]/[k\sqrt{2}]$$

$$q_i = s_i[\sqrt{k+1} - 1]/[k\sqrt{2}]$$

The first of these quantities, p_i, represents a large fraction of the step size. The other quantity, q_i, represents a small fraction of the step size. These quantities are added to the coordinates of the starting vertex as shown in Table 6.4.

At first glance, Table 6.4 appears to be highly complex, but in fact there is considerable simplifying structure in it. For example, the first row simply represents the starting vertex with coordinates $S_1, S_2, S_3, \ldots, S_k$. All other initial vertexes originate from this starting vertex.

If the first row of Table 6.4 is ignored, the remaining rows and columns form a $k \times k$ square matrix. Note that values of p_i are added to values of S_i to form the main diagonal elements of this matrix (extending from upper left to lower right). The remaining off-diagonal elements of this matrix are formed by adding values of q_i to values of S_i. Because the p_i values represent a large fraction of the step size and the q_i values represent a small fraction of the step size, each successive initial vertex (after the first) represents a large excursion in a particular factor dimension.

The geometry of this tilted initial simplex is shown in Figure 6.4. The starting vertex is shown at (10.00, 10.00). Step sizes of 80.00 were specified for each factor. Thus, for each factor, $p_i = 77.27$ and $q_i = 20.71$. Adding these p_i and q_i values to the starting vertex as shown in Table 6.4 gives vertexes at (87.27, 30.71) and (30.71, 87.27).

When all factors are measured in the same metric and when the step sizes are the same for each factor, the tilted algorithm produces an equilateral simplex (e.g., the equilateral triangle in Figure 6.4). In practice, however, the tilted initial simplex algorithm rarely produces equilateral simplexes. This is because the metrics are seldom the same for each factor (unless all factors have been coded or scaled) and because the step sizes are seldom the same.

The tilted initial simplex algorithm is mathematically elegant and looks good on paper. The tilted initial simplex algorithm is not known to have any practical advantage over the corner algorithm, discussed next.

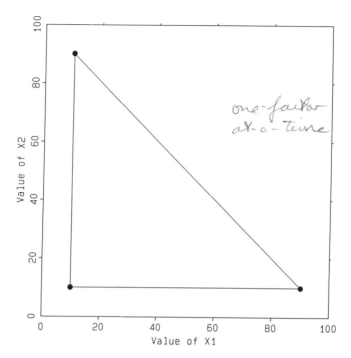

Figure 6.5 A two-factor simplex generated by the corner initial simplex algorithm.

Corner Initial Simplex

In Table 6.4, if each $p_i = s_i$ and each $q_i = 0$, then Table 6.5 results. This is the tabular representation of the $k + 1$ vertexes in the corner initial simplex algorithm. The effect of this algorithm is shown in Figure 6.5. In this example, as in Figure 6.4, the initial vertex is at (10.00, 10.00) and the step size is 80.00 in each factor. However, because all other factors are held constant at their S_i values while one factor at a time is incremented an amount equal to s_i, the result is an orthogonal (right-angled) arrangement of vertexes that is equivalent to changing only one factor at a time. Thus, the additional vertexes lie at (90.00, 10.00) and (10.00, 90.00). Experimental designs that have been generated by this "one-factor-at-a-time" method are also known as the first-order Koshal designs [14, 15].

The corner initial simplex algorithm derives its name from the fact that the simplex produced by this method can be translated to the origin and will fit into the corner defined by the factor axes.

If the corner initial simplex algorithm is used with the variable-size simplex, the orthogonality of vertex coordinates is lost after only a few moves. This is illustrated in Figure 6.6.

If a corner initial simplex is used with the fixed-size simplex algorithm and only two factors, then the orthogonality of vertex coordinates is retained. This is illustrated in Figure 6.7. The simplex was started at (80.00, 80.00) with step sizes of 10.00 in each factor. The simplex converged at (20.00, 40.00). In higher

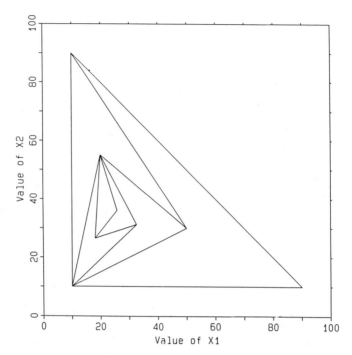

Figure 6.6 An illustration of the loss of orthogonality of the initial corner simplex when used with the variable-size simplex algorithm.

dimensional factor space, subsequent simplexes produced from a corner initial simplex do not usually possess orthogonality of vertex coordinates (see, for example, Figure 6.18).

PLACEMENT AND SIZE
OF THE INITIAL SIMPLEX

The size and placement of the initial simplex often depend on the environment in which the simplex is used. The following considerations generally apply.

Manufacturing

In manufacturing, systems for making products are already in operation. It is unwise to make large changes in process factors: the risk of producing nonconforming material is too great. If simplex optimization is to be employed in manufacturing, it should be done in the spirit of Box's original concept [16] (see Chapter 2): very small changes in each factor, with vertexes evaluated continuously until enough statistical power can be achieved to decide with a given level of confidence which vertex should be rejected.

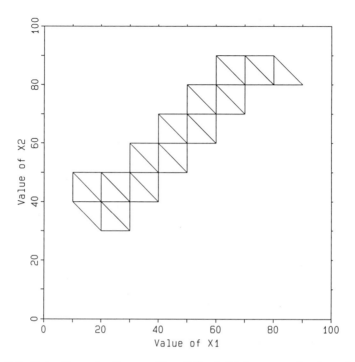

Figure 6.7 Retention of orthogonality of the initial corner simplex when used with the fixed-size simplex algorithm and only two factors.

Thus, in manufacturing, small simplexes are usually desirable. To be certain that small simplexes do not become large simplexes, the fixed-size sequential simplex algorithm is usually used.

Research

In research, rapid achievement of promising factor combinations is required. Although safety is always a concern, the consequences of large changes in factors are usually less severe on the small scale of research than on the large scale of manufacturing. If simplex optimization is to be employed in research, it should probably be done in a bold manner: very large changes in each factor to scout the factor space followed by later vertexes that better define the region of the optimum.

Thus, in research, large simplexes are usually desirable. But to be certain they eventually become small simplexes, the variable-size sequential simplex algorithm is usually used.

Development

Product or process development is a domain somewhere between research and manufacturing. Investigators in research usually claim to carry out "research and

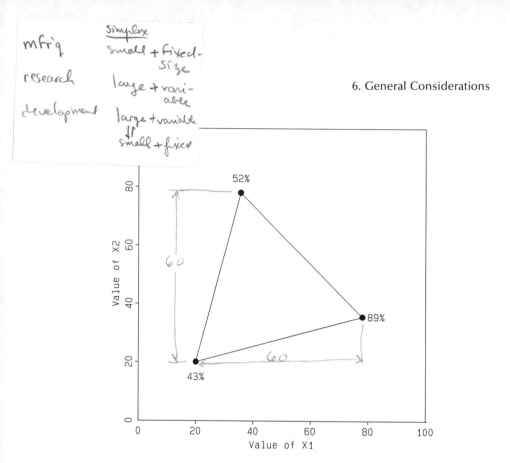

Figure 6.8 A large, two-factor, tilted initial simplex. The numbers beside the vertexes indicate response.

development." Investigators in manufacturing often take issue with this, and claim that they (the manufacturers) carry out "development and manufacturing." It is probably safe to say that workers in research do some amount of development, but workers in manufacturing do the remaining amount of development necessary to make a product manufacturable. This gray area called "development," lying between research and manufacturing, involving communication between researchers and manufacturers, is often poorly defined with blurred boundaries. It is an area of vital concern to managers concerned about quality (see Chapter 1).

Because of the ambiguous nature of development, it is not possible to make general statements about which simplexes are appropriate. In general, though, large variable-size simplexes might be used if the development is still close to research. If the development is more advanced and represents a situation closer to manufacturing, the smaller, fixed-size simplexes might be preferred.

EXAMPLES OF INITIAL SIMPLEXES

Figure 6.8 is an example of a large, two-factor, tilted initial simplex. The starting vertex is (10.00, 10.00) with step sizes of 60.00 in each factor. The simplex spans the bounded factor space well.

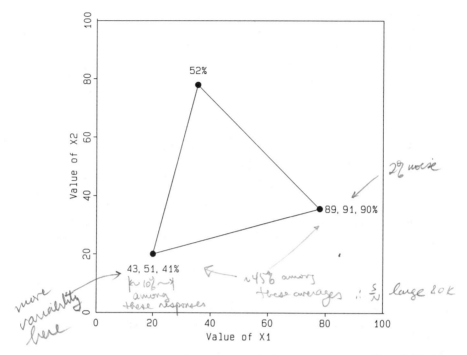

The handwritten annotations on the figure read: "more variability here", "for 10% among these responses", "~45% among these averages", "$\frac{5}{2}$ large ror", "2% noise".

Figure 6.9 The results of replicate experiments at (20.00, 20.00) and (77.96, 35.53).

In this example (Figure 6.8), the response is measured in percent. Values of 43, 52, and 89% have been obtained. Assuming the objective is to maximize the response, the value of 89% looks especially appealing. However, experimental data are always subject to uncertainty. One of the best ways to determine the magnitude of experimental uncertainty is to carry out replicate experiments. That is, to repeat experiments that have already been carried out.

Noise

Figure 6.9 shows the results of replicate experiments that were carried out at the vertexes (20.00, 20.00) and (77.96, 35.53). The repeatability at the vertex of lowest response is not very good (responses of 43, 51, and 41%), but the repeatability at the vertex of highest response is very tight (89, 91, 90%). This is typical of many response surfaces. Better responses occur near the tops of response surfaces where the slopes with respect to the factors are shallow: large variations in the factors produce little variations in the response. Poorer responses occur along the sides of response surfaces where the slopes with respect to the factors are steep: large variations in the factors produce large variations in the response.

sd of replicates
——————————————
sd of vertex responses

It is important to know how much purely experimental uncertainty there is in a system before the sequential simplex method is used. ("Pure error" is the name statisticians give to purely experimental uncertainty.) Large amounts of uncertainty in the response will cause the simplex to wander erratically. A rough measure of "large" in this context is the ratio between the standard deviation of replicate measurements (the square root of the "pure error variance") and the standard deviation of the vertex responses (or their averages if replicates have been carried out).

If it is discovered that there is excessive uncertainty in the response, steps should be taken to bring the system under better statistical control–that is, to reduce the uncertainty by tightening the control of all factors, including those that are explicitly included in the optimization as well as those that are supposedly being held constant and not varied. Nested [17, 18] and staggered nested [19] experimental designs are particularly useful in this context. If the uncertainty can be brought under control, then the simplex will work well.

If uncertainty cannot be decreased to a satisfactory level, the simplex can still be used if each vertex is repeated a number of times to average out the uncertainty. In a manufacturing environment where small, fixed-size simplexes are found and where relatively large (as defined above) amounts of uncertainty are typically encountered, this might be appealing–after all, "experiments" will be carried out whether or not there is any intentional variation of the factors. It was this manufacturing environment in which Box put forth his original insight into evolutionary operation [16], and the environment in which Spendley, Hext, and Himsworth originally introduced the sequential simplex [12].

In a research environment where large, variable-size simplexes are found and where relatively large amounts of uncertainty are occasionally found, the simplex might be run for a while anyway. Chances are that if the initial simplex is large, the vertexes will produce a set of responses that has greater variability than the variability due to purely experimental uncertainty. Such a situation is shown in Figure 6.9. Although replicate responses of 43, 51, and 41% have a range of about 10% (absolute), the range of average responses among vertexes is about 45% (absolute). Thus, although the system is rather noisy, the "signal-to-noise ratio" is still large enough that correct algorithmic decisions will be made much of the time.

Finally, there is also the alternative of not using the sequential simplex method but using another type of experimental design instead. Classical factorial-type designs have built-in noise-reducing capabilities. Central composite designs [20] and Box–Behnken designs [21] are very effective, especially if replication can be included and if the number of factors is relatively small (usually six or fewer).

We recommend the following:

rule

> After the initial set of experiments has been carried out at the $k + 1$ initial simplex vertexes, DO NOT immediately do a reflection. Instead, repeat the vertex that gave the worst response and repeat the vertex that gave the best response, two or three times each, if possible.

This will provide some idea of how much uncertainty can be expected from the system. A decision can then be made as to whether to proceed with the simplex algorithm.

Alternatively (and often more efficiently) *any* vertex can be repeated before any of the other k vertexes are evaluated. This will give an immediate indication of the amount of noise that might be expected from the system.

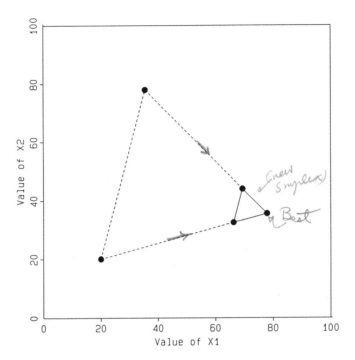

Figure 6.10 Construction of a second "initial" simplex by retaining the vertex giving the best response and moving all remaining vertexes toward the best vertex.

Quitting after the Initial Simplex

Returning to the example of Figures 6.8 and 6.9, the response "percent" is naturally bounded between 0 and 100%. These are known reference values for the concepts "perfectly good" and "perfectly bad." Thus, conclusive statements can be made about the observed responses of 43, 52, and 89%. In particular, a response of 89% out of a possible 100% looks very good, especially when compared with 43 and 52%. Even in the presence of approximately 10% (absolute) uncertainty in the response, it is clear that the vertex giving 89% is much closer to a perfect response than are the vertexes giving the other responses.

If this large, variable-size simplex is allowed to continue, it will eventually collapse to the region around the best vertex. But as it collapses, it will probably produce many responses that are not as good as 89%. Thus, rather than spend time and effort letting the simplex collapse by itself, it is possible to stop the current simplex and construct a new, smaller simplex in the region of the currently best vertex.

One possibility for such a contraction is to retain the vertex giving the best response and move all of the remaining vertexes toward the best vertex as shown in Figure 6.10. This requires evaluating k new responses. A second possibility is shown in Figure 6.11. In this case, the new simplex straddles the previous best vertex and will require the evaluation of $k + 1$ new responses. Other possibilities exist:

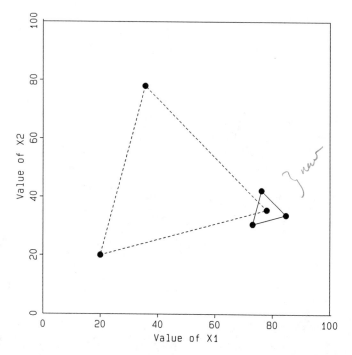

Figure 6.11 Construction of a second "initial" simplex by moving all vertexes toward the best vertex and then translating the smaller simplex so it is centered on the previous best vertex.

for example, using the previous best vertex as the starting vertex for the new simplex and specifying small step sizes in each factor.

Centered Thirds

One intriguing possibility for placing the initial simplex is to center it at the midpoint of all factor domains and use a step size equal to one-third of the domain in each factor. An example is shown in Figure 6.12. Using this strategy, the simplex can move quickly toward the boundaries of the factor domain.

BOUNDARY VIOLATIONS

As suggested by Figure 6.13, the simplex algorithms will frequently ask the investigator to carry out an experiment that is outside the factor boundaries. Boundary violations can be handled in a number of ways. In all of the following examples, it is assumed that the boundaries are simple single-factor boundaries—that is, the feasible region of factor space lies within a k-dimensional hypercube.

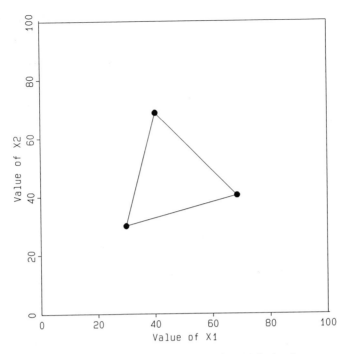

Figure 6.12 The "centered thirds" initial simplex.

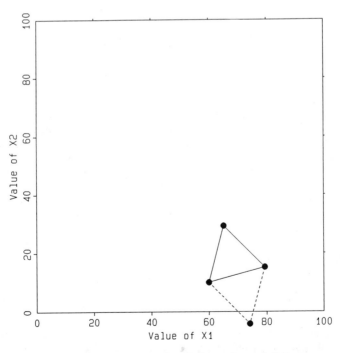

Figure 6.13 An example of a boundary violation.

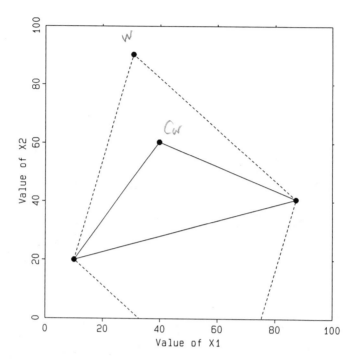

Figure 6.14   Assigning a very undesirable response to the out-of-bounds reflection vertex forces the variable-size simplex to carry out a C_W contraction.

Remove the Boundary

Perhaps the simplex is trying to tell the investigator that the optimum lies beyond the boundary. If so, consider removing the boundary constraint. Sometimes boundaries are set for convenience and are not absolute. For example, if a low temperature boundary of $0\,°C$ is necessary simply because a water-bath is used to control the temperature, try using some other solvent (e.g., ethanol) instead. If the water-bath dictates a high temperature boundary of $100\,°C$, try using an oil-bath instead.

Sometimes a bit of imagination is required. For example, if the simplex suggests adding -0.103 mL of $0.10\ M$ HCl, try adding $+0.103$ mL of $0.10\ M$ NaOH instead.

In some cases, boundaries will have been set because of safety considerations. These boundaries should not be removed.

Variable-Size Simplex

If you cannot remove the boundary, then you cannot do the experiment. In the case of the variable-size simplex algorithm, simply assign the out-of-bounds vertex an infinitely bad response. This will automatically force a C_W contraction and the simplex will pull back within bounds (see Figure 6.14).

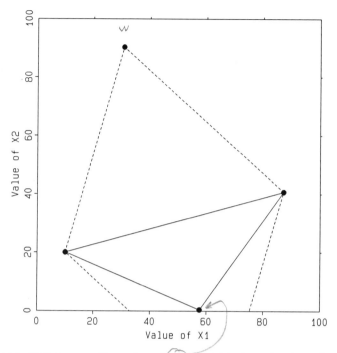

Figure 6.15 If **R** is out of bounds and C_R is within bounds, then C_R might be carried out instead of C_W.

If a boundary violation occurs with the variable-size simplex, some investigators will first test to see if the C_R vertex lies within bounds (see Figure 6.15). If C_R does lie within bounds, then they will retain this vertex instead of collapsing back to C_W. The rationale is that because the simplex wanted to move away from **W**, carrying out C_R if possible is more in keeping with the spirit of the algorithm than is carrying out C_W. Simply carrying out C_W because of a boundary violation goes against the intent of the original algorithm.

Fixed-Size Algorithm

In the case of the fixed-size simplex algorithm, assign the out-of-bounds vertex a previously determined, numerically undesirable response *plus the vertex number* (e.g., $-1,000,000 + 7$; the sum should also be undesirable). This way, the older vertexes will look worse and be rejected before the more recent vertexes. The simplex will then spin back within bounds. An example of this is shown in Figure 6.16. The numbers within the simplexes show the sequence of moves.

The two vertexes that are out of bounds in Figure 6.16 represent "phantom experiments": even though coordinates were calculated for these vertexes, no experimental effort was required there.

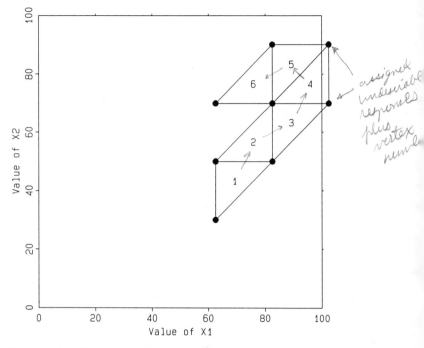

Figure 6.16 Behavior of the fixed-size simplex at a boundary.

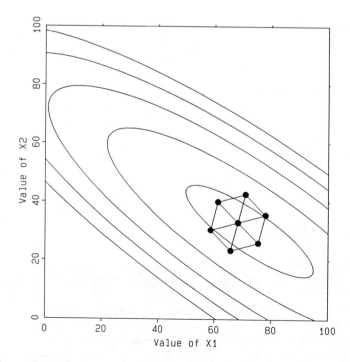

Figure 6.17 Convergence of the fixed-size simplex in two dimensions.

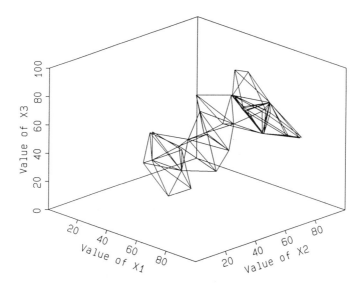

Figure 6.18 A three-factor simplex moving in factor space. The initial corner simplex does not retain orthogonality in higher dimensional factor space. Thirty-five simplexes are shown.

CONVERGENCE

At some point, the sequential simplex must be halted. Over the years, many mathematical convergence criteria have been proposed. Some are based on changes in response. Others are based on changes in factors. Most look good on paper, in noise-free environments, but many fail in real-life situations. We present here some examples of convergence, and discuss three practical convergence criteria.

The Fixed-Size Simplex

Figure 6.17 shows a two-factor, fixed-size simplex circling in the region of the optimum. Determination of convergence of the fixed-size simplex in two dimensions is simple: when two successive new reflection vertexes overlap previous vertexes, the simplex will circle and has converged.

It is necessary that more than one successive new reflection vertexes overlap. Otherwise the simplex could break out and move in a different direction. Examples of "circled" simplexes breaking out in new directions were seen in Figures 5.7, 5.11, and 5.15.

In higher dimensional factor space, the fixed-size simplex will not form the higher dimensional analogs of the circle: the sphere and hypersphere. Instead, the

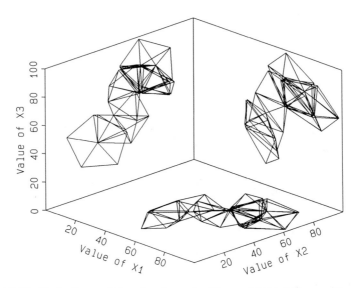

Figure 6.19 Projections of the simplexes in Figure 6.18 into lower dimensional factor spaces.

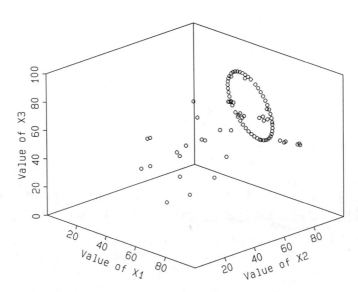

Figure 6.20 Vertexes of a three-factor simplex moving in factor space showing coning of the converged simplex. Eighty-eight vertexes are shown.

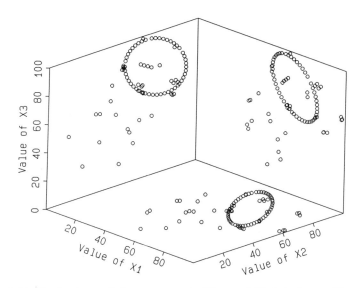

Figure 6.21 Projections of the vertexes in Figure 6.20 into lower dimensional factor spaces.

simplex becomes anchored to a best vertex and spins in a cone or hypercone about this best vertex, producing a lower dimensional pattern of responses at the suboptimal vertexes.

Figure 6.18 shows a three-dimensional fixed-size simplex moving through factor space. The initial simplex was constructed about a starting vertex of (35.00, 35.00, 35.00) with a step size of 20.00 in each factor using the corner algorithm. The optimum in the response surface is located at (80.00, 80.00, 80.00). As seen in Figure 6.18, the fixed-size simplex moves from the starting vertex toward the region of the optimum. Thirty-five simplexes (38 vertexes) are shown.

Figure 6.19 shows two-factor projections of the three-dimensional simplex of Figure 6.18. The images of the simplexes are projected down into the x_1–x_2 plane; left into the x_2–x_3 plane; and back into the x_1–x_3 plane. Again, 35 simplexes (38 vertexes) are shown.

Figure 6.20 is a representation that simply plots the locations of each vertex with an open circle symbol and omits the lines joining the vertexes. Again, the simplex is started at (35.00, 35.00, 35.00) with a step size of 20.00 in each factor using the corner algorithm, but Figure 6.20 shows 85 simplexes (88 vertexes). The coning of the converged simplex is readily apparent from this figure. Figure 6.21 shows the two-factor projections of Figure 6.20.

Usually, a fixed-size simplex would never be run long enough to observe coning in an experimental system. There would normally be no reason to carry out so many experiments to verify that the region of the optimum had been located. However, there is one situation where the continued use of a small fixed-size simplex

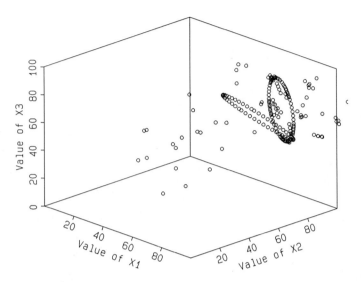

Figure 6.22   Vertexes of a three-factor simplex moving toward a shifting optimum showing coning of the converging simplexes. Two-hundred and three vertexes are shown.

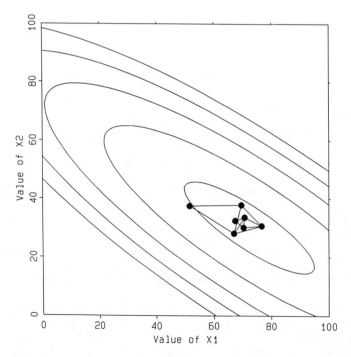

Figure 6.23   Convergence of the variable-size simplex in two dimensions.

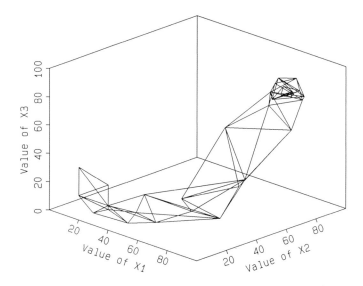

Figure 6.24 A small corner simplex converging to an optimum.

might have some economic advantage, and that is a manufacturing setting in which the location of the optimum drifts slowly.

In Figure 6.22, 200 simplexes (203 vertexes) are shown. The optimum in factor x_1 drifts slightly between successive simplexes; otherwise, the behavior of the system is as that shown in Figure 6.20. Note that the initial coning is at a location different from that for the nondrifting system. But after circling in one area for some time, the simplex finds a better response and cones around that vertex. Further drift would lead to coning in a different location.

The phenomenon of coning deserves further attention, from both a mathematical point of view as well as a practical point of view. The ability to follow drifting optima is a matter of concern for modern statistical process control.

The Variable-Size Simplex

Figure 6.23 shows a variable-size simplex converging in two dimensions. As expected, the simplex contracts onto the optimum.

Figure 6.24 shows an initially small, three-factor, variable-size simplex traveling across the factor space and converging on an optimum. In this example, the initial simplex was constructed with a starting vertex at (10.00, 10.00, 10.00) and a step size of 20.00 in each factor using the corner algorithm. The two-factor projections are shown in Figure 6.25. Note that the orthogonality of the initial corner simplex is quickly lost.

Figure 6.26 shows an initially large, three-factor, variable-size simplex traveling across the factor space and converging on an optimum. In this example, the initial simplex was constructed with a starting vertex at (10.00, 10.00, 10.00) and a

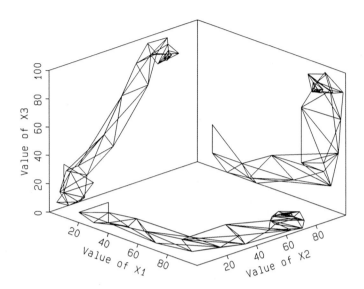

Figure 6.25 Projections of the simplexes in Figure 6.24 into lower dimensional factor spaces.

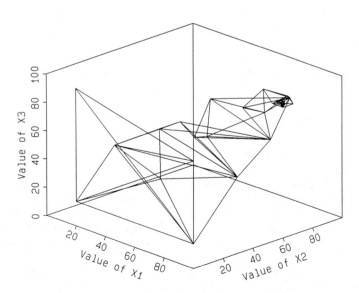

Figure 6.26 A large corner simplex converging to an optimum.

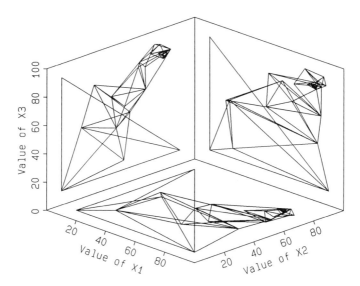

Figure 6.27 Projections of the simplexes in Figure 6.26 into lower dimensional factor spaces.

step size of 80.00 in each factor using the corner algorithm. The two-factor projections are shown in Figure 6.27. Again, note that the orthogonality of the initial corner simplex is quickly lost.

Threshold Criterion

In most optimization projects, there is one simple convergence criterion: stop when you get what you need. This threshold criterion is illustrated in Figure 6.28.

In many projects, especially at the early research stage but also at the later development stage, certain "benchmark" performance criteria must be achieved before the project is allowed to go on to other stages of development and implementation. In the pharmaceutical industry, for example, it is often necessary to demonstrate a minimum achievable synthetic yield before a candidate molecule can begin the lengthy trials and evaluations required to get it on the market.

In such cases, the threshold is usually well defined and the goal of the work is to achieve that threshold. At this stage, being able to do better than the threshold is not important. Perhaps it will become important later, but it is not now. If that is the case, then stop the simplex when the threshold has been achieved.

Budget Criterion

Most projects have a budget. Box, Hunter, and Hunter [22] suggested that the experimental design part of the budget should be partitioned 25% for scouting

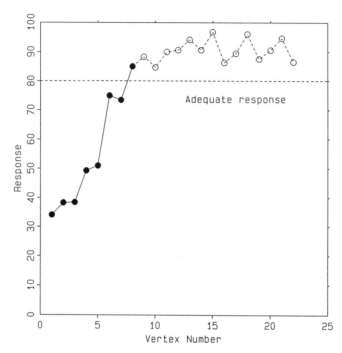

Figure 6.28 The convergence criterion of adequate response. After the threshold is reached, further experimentation is unjustified.

experiments, 50% for goal achievement, and the remaining 25% for exploring profitable possibilities suggested by results from the first 75% of the budget. Presumably, the simplex would be used in the first 25% or middle 50% of the expenditure schedule.

If the simplex has not achieved the desired results by the time the budget is spent, do not spend any more money on that system. Perhaps the simplex is trying to tell you that the system is not capable of the performance you desire. There is nothing magic about the simplex.

Noise Criterion

In the presence of noise, the simplex will begin to wander if the differences between responses among the simplex vertex are on the same order as the purely experimental uncertainty in the response. If (as was suggested before) the best and worst vertexes were replicated before carrying out the first reflection, there will be an estimate of the purely experimental uncertainty. Periodic monitoring of the variation among vertex responses will provide some indication if noise is becoming a problem.

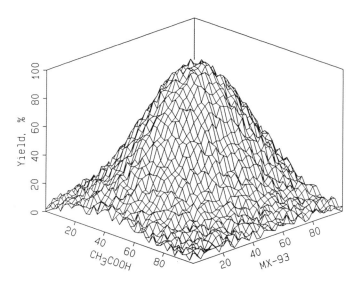

Figure 6.29 A realistic view of an experimental response surface.

THE *k* + 1 RULE

Figure 6.29 gives a realistic view of a response surface. Because of uncertainties in the performance of the system and because of uncertainties introduced by the measurement process, the response surface is noisy, "fuzzy," and uncertain. Nonetheless, the simplex must try to move toward the optimum in the presence of this noise. Two situations are of special interest.

An Incorrectly Poor Response

Occasionally, a response will be recorded that is incorrectly poorer than the underlying response surface. This might happen, for example, if the experimenter transposes two digits in the numerical representation of the response. Or perhaps a measuring instrument was not calibrated properly. Or a mistake was made in the setting of a factor. Whatever the reason, the response is not representative of the underlying response surface – it appears that the response is worse than it really is.

On a simplex worksheet this new, incorrectly poor response will go into the row labeled **N** and be rejected on the next simplex iteration. As shown in Figure 6.30, the simplex might be thrown off track for a few moves, but no lasting harm will be done. Incorrectly poor responses are quickly eliminated from the simplex.

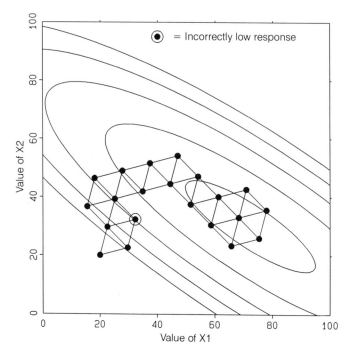

Figure 6.30 Illustration of how an incorrectly poor response throws the fixed-size simplex temporarily off course.

An Incorrectly Good Response

Occasionally, responses will be recorded that are incorrectly better than the underlying response surface. Again, there are any number of reasons why this might occur.

On a simplex worksheet this new, incorrectly good response will go into the row labeled **B** and will be retained on the next simplex iteration. In fact, if it looks good enough, it will be retained for all future simplex iterations. Thus, as shown in Figure 6.31, there is a real danger that the simplex will become fastened to this incorrectly good response and converge around a false optimum.

Suspiciously Good Responses

Because of the possibility of recording an incorrectly good response, Spendley, Hext, and Himsworth [12] proposed that if a vertex were in a simplex $k + 1$ times and was not about to be rejected, the progress of the simplex should be halted and that vertex reevaluated. This is known as the "$k + 1$ rule."

The possibility of an incorrectly good response is why the "times retained" column on the simplex worksheets is incremented each time a vertex is transferred

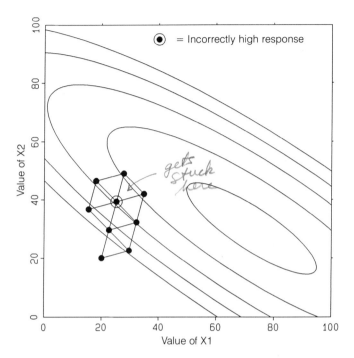

Figure 6.31 Illustration of how an incorrectly good response causes the simplex to converge on a false optimum.

to a new worksheet. It allows the experimenter to monitor how long a particular vertex has remained in a simplex. If that number is *k* + 1, and if the vertex is not about to be rejected, then the vertex should be reevaluated.

The reevaluation can be used in any one of a number of ways. Some workers prefer to replace the original response with the reevaluated response; the philosophy here is that the most recent response is the most representative response, especially if the system is drifting slightly. Other workers prefer to average the original response with the reevaluated response; the philosophy in this case is that the mean of the two evaluations is a better estimate of the response than any one individual evaluation. A combination of these two philosophies is behind the use of weighted averages, where the most recent evaluation receives the most weight. (Usually the simplex will be halted before many *k* + 1 rule reevaluations are carried out for any given vertex.)

There is no firm theoretical basis for the use of the value *k* + 1 for this rule. In our experience, the use of *k* + 1 is overly conservative. The value *k* + 3 seems to be more satisfactory in practice.

If the sequential simplex is being used for numerical optimization (e.g., minimizing the sum of squares of residuals for a nonlinear model), there is no need to use the *k* + 1 rule.

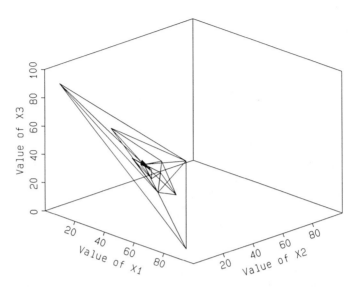

Figure 6.32 A degenerate three-factor simplex used to optimize a constrained three-component mixture. See text for details.

MIXTURE DESIGNS

In formulations work, there is a constraint that the sum of the component percentages must add up to 100%. This is known as the mixture constraint. Thus, if the formulation has three components, only two of the components are independent: the third component must be such that the sum adds up to 100%.

Special experimental designs can be constructed for formulations or mixture experiments. These are known as *simplex mixture designs*, and are different from the sequential simplex discussed in this book. Some of the simplex mixture designs can be constructed from constrained full factorial designs. The book by Cornell is recommended [23].

The sequential simplex discussed in this book can be used to optimize mixtures also. One way to optimize a three-component mixture is to define a factor space in two of the three factors (say x_1 and x_2), constrain that factor space so that $(x_1 + x_2) < 100\%$ (this will be a diagonal boundary across the factor space), and let $x_3 = 100\% - x_1 - x_2$. The sequential simplex can then be used to determine vertexes in x_1 and x_2, and x_3 can be calculated by difference.

There is also a way to carry out mixture experiments by forcing the simplex to move in a constrained three-dimensional factor space. This is shown in Figure 6.32. Neither the tilted nor corner algorithm was used here. Instead, a user-supplied starting simplex was required. All vertexes are such that the sum of the factors adds up to 100%. The initial vertexes are (5.00, 5.00, 90.00), (5.00, 90.00, 5.00), (90.00, 5.00, 5.00), and (33.33, 33.33, 33.33). These initial vertexes are the three extreme vertexes in the oblique triangle and its center point in Figure 6.32. This degenerate tetrahedron (i.e., a triangular simplex with center point) exists in

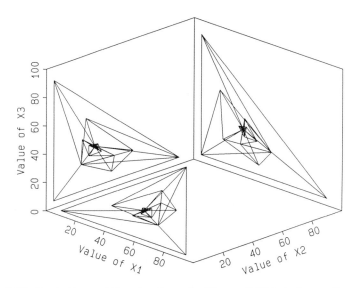

Figure 6.33 Projections of the simplexes in Figure 6.32 into lower dimensional factor spaces.

a three-dimensional factor space but is itself a two-dimensional figure. In effect, the three-dimensional tetrahedron has been collapsed into an oblique two-dimensional plane.

After the initial simplex has been constructed, the remaining simplex moves follow the usual algorithmic pattern. The progress of the simplex is shown in Figure 6.32. Two-factor projections are shown in Figure 6.33. (The bounded triangular region projected into the x_1–x_2 plane is similar to the usual simplex-mixture presentation [23].)

STRATEGIES OF EXPERIMENTATION

The simplex method of optimization is a powerful yet simple technique for use in research, development, and manufacturing [11]. It is highly efficient, is capable of optimizing several factors at one time, and the variable-size simplex is not stranded by ridge systems. Interaction with the experimenter is not necessary: the simplex technique uses only empirical feedback and is therefore of special interest for investigations involving totally automated optimization.

Preliminary Screening Experiments

Most optimization schemes traditionally make use of a preliminary factorial design to determine those factors that have a significant effect on the response.

This step is especially important for optimization techniques that are themselves based on factorial designs. Such designs generally require m^k experiments in the initial pattern, where m is the number of levels and k is the number of factors to be investigated. Each move toward the optimum also requires approximately m^k experiments ($\frac{1}{2}m^k$ in the most favorable case) so that if g factorial designs are needed to reach the optimum, a total of approximately $g(m^k)$ experiments is required. Because this is often a very large number, those factors not found to be significant at some level of probability in the initial factorial experiment are usually excluded from investigation in the remaining factorial designs. The result is fewer factors (a smaller value for k) and a reduction in the total number of experiments.

There is a subtle but very important point involving significance testing, especially when used to reject factors in optimization studies. Tests of significance are based upon the *null hypothesis*, which states that a factor has *no* effect on the response. As Davies pointed out [24], "When a significant difference is found the experimenter is justified in asserting that the Null Hypothesis is probably false. When no significant difference is found he is not equally justified in asserting that the Null Hypothesis is true. . . . A real, and possibly important, difference may well have occurred, but the experiment may not have been sufficiently sensitive to detect it."

In selecting (for further investigation) factors that are significant at a given level of probability, the investigator is assured that those factors will probably be useful in improving the response. But this "efficiency" is gained at the expense of omitting factors that might also be significant.

Ideally, the question that should be asked is not, "What factors are significant at the p level of probability?" but rather, "What factors are *insignificant* at the p level of probability?" This involves an entirely different type of statistical testing. The number of factors retained when using the second criterion will in general be larger than the number retained when using the first; the investigator will, however, be assured that she is probably not omitting from investigation any factors that are important.

A second potential difficulty in using screening experiments to select factors for optimization is that if a factor is continuous, then the results are highly dependent on the choice of levels tested. This is especially true in the case of two-level factorial designs (e.g., Plackett-Burman [3], saturated fractional factorial [1, 2], Hadamard [4–6], Taguchi [7–9]) in which no information about second-order effects (curvature) can be obtained. If the levels of a factor are by chance chosen on either side of the optimum in such a way that they give essentially identical response, then the results of the screening experiments will not show that the factor is significant at all. Factor effects that are apparently not significant can also be obtained for significant factors if the levels of the factor are chosen too close together (anywhere in the factor domain, whether curvature is present or not).

In view of these limitations of screening designs and hypothesis testing, separate, preliminary screening experiments are possibly unnecessary prior to simplex optimization.   *Do simplex before screening.*

The Efficiency of the Sequential Simplex

Two important characteristics of the simplex method are the small number of experiments required for the initial simplex ($k + 1$) and the efficiency of finding the

But, what about centerpoints?

k = # of factors
g = # of factorials

optimum. Including an extra factor in the simplex search adds only one more experiment to the initial design. The number of experiments required to reach the region of the optimum is approximately $g + k + 1$ for the fixed-size simplex, and at most $2 \times g + k + 1$ for the variable-size simplex. Although adding more factors does increase somewhat the number of steps necessary to reach the optimum (g) using the simplex method, this number is still relatively small.

A Strategy for Research, Development, and Manufacturing

Statistics has often been used to help answer, in order, three questions commonly encountered in research, development, and manufacturing [25]:

1. Does an experimentally measured response depend on certain factors?
2. What equation does the dependence best fit?
3. What are the optimum levels of the important factors?

As Driver [25] has pointed out, "The questions are so related that it is possible for the experimenter not to know which one he wishes to answer and, in particular, many people try to answer question 2 when in fact they need the answer to the narrower question 3."

When optimization is the desired goal, questions of significant factors and functional relationships are usually of interest only in the area of the optimum. The simplex method offers a means of rapidly and efficiently finding the region of the optimum, *after which* the functional relationship and then the significant factors can be determined. We feel that in many cases statistics should be used to help answer, in reverse order, the three questions commonly encountered in research, development, and manufacturing:

3. What are the optimum levels of the important factors?
2. What equation does the dependence best fit?
1. Does an experimentally measured response depend on certain factors?

This suggests first using a rapid optimization method such as the sequential simplex, followed by classical screening designs and response-surface mapping designs.

order
1. Simplex
2. classical screening
3. response-surface mapping

Simplex Fanaticism

As Brian Joiner has stated [26], "When the only tool you have is a hammer, it's amazing how many problems begin to look like nails." In the past, many statisticians have probably been guilty of trying to solve every experimental design problem with the factorial design hammer. More investigators (and statisticians) need to be more aware of the variety of statistical experimental design tools available to them.

Simplex is one of many statistical tools the researcher should have in her toolkit. But not every problem can (or should) be solved with the sequential simplex method. If the problem involves unrankable, truly discrete factors, if the problem has excessive noise, or if the problem requires a long time between experiments, then experimental designs other than the sequential simplex might be more appropriate.

As suggested in Chapter 3, there is a temptation to become fascinated by the movement of the simplex, to get caught up in the mathematics and logic of the simplex moves. There is often a desire to do yet one more vertex, just to see if the response improves further. This is usually unwise, and is wasteful of experimental resources.

In an experimental environment, the simplex becomes less and less efficient as it approaches the region of the optimum (see Chapter 5). Simplex optimization is most effective when it is used to get to the *region* of the optimum as rapidly as possible. Then, once in the region of the optimum, classical experimental designs should be used to investigate the effects of the factors and to describe the behavior of the system [*11*].

REFERENCES

1. G. E. P. Box and J. S. Hunter, "The 2^{k-p} fractional factorial designs, Part I," *Technometrics*, **3**, 311–351 (1961).

2. G. E. P. Box and J. S. Hunter, "The 2^{k-p} fractional factorial designs, Part II," *Technometrics*, **3**, 449–458 (1961).

3. R. L. Plackett and J. P. Burman, "The design of optimum multifactorial experiments," *Biometrika*, **33**, 305–325 (1946).

4. W. J. Diamond, *Practical Experiment Designs*, 2nd ed., Van Nostrand Reinhold, New York, 1989.

5. P. J. Ross, *Taguchi Techniques for Quality Engineering: Loss Function, Orthogonal Experiments, Parameter and Tolerance Design*, McGraw-Hill, New York, 1988.

6. D. J. Wheeler, *Tables of Screening Designs*, 2nd ed., Statistical Process Controls, Inc., Knoxville, TN, 1989.

7. A. Bendell, J. Disney, and W. A. Pridmore, Eds., *Taguchi Methods: Applications in World Industry*, IFS Publications, Springer-Verlag, London, 1989.

8. G. Taguchi, *Introduction to Quality Engineering: Designing Quality into Products and Processes*, Kraus International Publications, White Plains, NY, 1986.

9. D. J. Wheeler, *Understanding Industrial Experimentation*, Statistical Process Controls, Inc., Knoxville, TN, 1987.

10. W. K. Dean, K. J. Heald, and S. N. Deming, "Simplex optimization of reaction yields," *Science*, **189**, 805–806 (1975).

11. S. L. Morgan and S. N. Deming, "Simplex optimization of analytical chemical methods," *Anal. Chem.*, **46**, 1170–1181 (1974).

12. W. Spendley, G. R. Hext, and F. R. Himsworth, "Sequential application of simplex designs in optimisation and evolutionary operation," *Technometrics*, **4**, 441–461 (1962).

13. Gordon S. G. Beveridge, and Robert S. Schechter, *Optimization: Theory and Practice*, McGraw-Hill, New York, 1970.

14. G. E. P. Box and N. R. Draper, *Empirical Model-Building and Response Surfaces*, Wiley, New York, 1987.

15. R. S. Koshal, "Application of the method of maximum likelihood to the improvement of curves fitted by the method of moments," *J. R. Stat. Soc.*, **A96**, 303–313 (1933).

16. G. E. P. Box, "Evolutionary operation: A method for increasing industrial productivity," *Appl. Stat.*, **6**, 81–101 (1957).

17. R. E. Lund, "Plans for blocking and fractions of nested cube designs," *Commun. Stat.-Theor. Meth.*, **11**, 2287–2296 (1982).

18. R. E. Lund, "Description and evaluation of a nested cube experimental design," *Commun. Stat.-Theor. Meth.*, **11**, 2297–2313 (1982).

19. J. R. Smith and J. M. Beverly, "The use and analysis of staggered nested factorial designs," *J. Qual. Technol.*, **13**, 166–173 (1981).

20. G. E. P. Box and K. B. Wilson, "On the experimental attainment of optimum conditions," *J. R. Stat. Soc.*, **13**, 1–45 (1951).

21. G. E. P. Box and D. W. Behnken, "Simplex-sum designs: A class of second order rotatable designs derivable from those of first order," *Ann. Math. Stat.*, **31**, 838–864 (1960).

22. G. E. P. Box, W. G. Hunter, and J. S. Hunter, *Statistics for Experimenters: An Introduction to Design, Data Analysis, and Model Building*, Wiley, New York, 1978.

23. J. Cornell, *Experiments with Mixtures: Designs, Models, and the Analysis of Mixture Data*, 2nd ed., Wiley, New York, 1990.

24. O. L. Davies, Ed., *The Design and Analysis of Industrial Experiments*, 2nd ed., Hafner, New York, 1963.

25. R. M. Driver, "Statistical methods and the chemist," *Chem. Br.*, **6**, 154–158 (1970).

26. B. L. Joiner, comment at the "[W. E.] Deming Conference for Statisticians," New York University, New York, 28–30 March 1988.

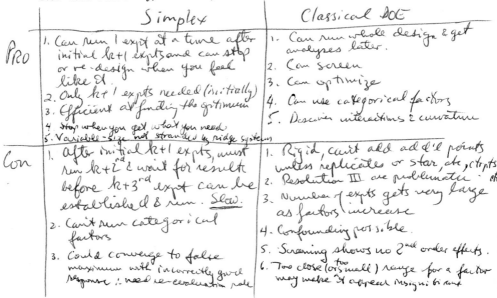

Chapter 7

Additional Concerns and Topics

SOME MINOR TOPICS

The following are a number of minor topics related to optimization and response surfaces.

Point-through-Point Reflection

Throughout this text we have emphasized that reflections are point-through-point reflections, not mirror-image reflections. The difference between the two types of reflections is shown in Figure 7.1.

When the simplexes are fixed-size and equilateral, the two operations are equivalent. In all other cases, the two operations are not equivalent.

Mirror-image reflection of nonequilateral simplexes is often not well defined. For example, in the fixed-size simplex of Figure 7.1A the next move would require reflection across a virtual surface (i.e., the extension of the line segment defining one face of the new simplex). In contrast, point-through-remaining-centroid reflection is always well defined (see Figure 7.1B).

Optima That Are Minima

Most of the examples in this text have shown the optimum as a maximum. Figure 7.2 is an example of an optimum that is a minimum. Such optima might correspond

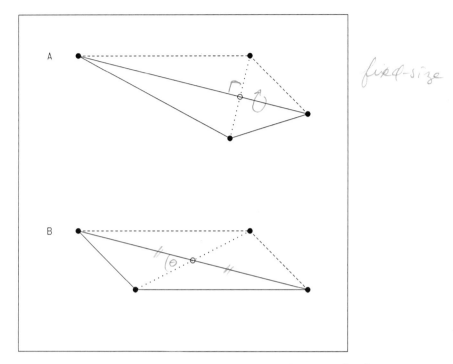

Figure 7.1 (A) Mirror-image reflection. (B) Point-through-point reflection.

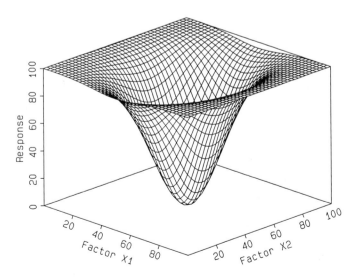

Figure 7.2 A response surface containing a minimum.

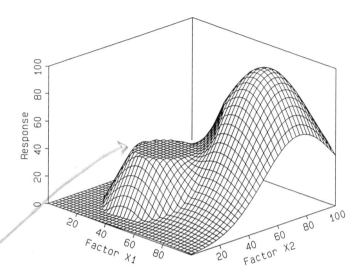

Figure 7.3 A response surface containing a plateau region.

to impurities in a manufactured drug as a function of throughput and pH, odor of a polymer as a function of initiator and temperature, etc.

Plateau Regions

Figure 7.3 shows a response surface with a plateau region near the center of the x_1–x_2 factor space. In this region, the response is relatively insensitive to changes in the factors. Although such regions might not produce optimal response (in this case, maximal response), they are nonetheless very useful from a quality control point of view. For reasons that should now be obvious, operating a process in a plateau region produces a statistical process control chart that is very tight. Such regions are not easily found by sequential simplex procedures – the response would have to include a measure of the gradient, which requires approximately $k + 1$ experiments *at each vertex*.

WHEN THE NEW VERTEX IS
THE WORST VERTEX

It sometimes happens that the new vertex will be the worst vertex. This situation occurred in Figures 3.18–3.20 and Figures 4.10–4.12, for example. This did not cause any problems with the progress of the simplexes, however. Both the fixed-size algorithm and the variable-size algorithm, as presented, allowed the simplex to move on and converge to the region of the optimum. In this section, we discuss in more detail what happens when the new vertex is the worst vertex.

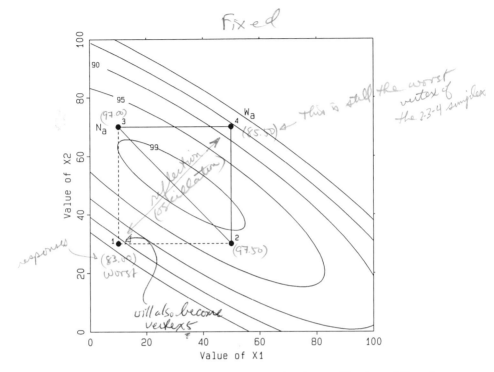

Figure 7.4 A fixed-size simplex stranded on a ridge by application of Spendley, Hext, and Himsworth's Rule 1 only. The new vertex is the worst vertex. N_a is the absolute next-to-the-worst vertex; W_a is the absolute worst vertex. See text for details. *(Vertex 4 will be reflected back to Vertex 1, in the 3rd simplex if only Rule 1 is applied) as vertex 5*

Fixed-Size Simplex

Figure 7.4 shows an example in which the new vertex is the worst vertex. The calculations for this move are shown in Figure 7.5. The solid lines in Figure 7.4 connect the vertexes in the new, current simplex (vertexes #2, #3, and #4). The dashed lines connect this current simplex with the vertex that was rejected in the previous simplex (vertex #1). The new vertex (vertex #4) has the worst response in the new simplex. The special labels N_a and W_a in Figure 7.4 refer to the absolute rankings of the responses in the current simplex, not to the rows of a worksheet.

Spendley, Hext, and Himsworth's "Rule 1" for implementing simplex EVOP [1] is to reject the vertex that has the worst response and reflect its coordinates through the centroid of the remaining hyperface an equal distance beyond to generate the reflection vertex. This is the rule that was used to generate the reflection vertex in Figures 7.4 and 7.5.

If Rule 1 were the only rule used to drive the fixed-size simplex toward an optimum, the simplex would eventually fail. This failure will occur for the simplex shown in Figure 7.4. Figure 7.6 gives the calculations that would be used to move the current simplex toward the next reflection vertex. Note that the

Simplex No. _1_ → _2_

Factor	X_1	X_2	Response	Rank	Vertex Number	Times Retained
Coordinates of	50.00	30.00	97.50	B	2	1
retained vertexes	10.00	70.00	97.00	N	3	1
Σ	60.00	100.00				
$\bar{P} = \Sigma/k$	30.00	50.00				
W	10.00	30.00	83.00	W	1	1
$(\bar{P} - W)$	20.00	20.00				
$R = \bar{P} + (\bar{P} - W)$	50.00	70.00	85.50	R	4	0

Figure 7.5 Calculations for generating vertex #4 in Figure 7.4. See text for details.

vertexes are ranked in absolute order from best (**B**), through next-to-the-worst ($\mathbf{N_a}$), to worst ($\mathbf{W_a}$).

The resulting reflection vertex (vertex #5) has the same coordinates as the previously rejected vertex (vertex #1)! If an experiment is carried out at the coordinates of vertex #5 (that is, at the coordinates of vertex #1 again), the response will probably be about the same as the response obtained initially at this factor combi-

Simplex No. _2_ → _3_

Factor	X_1	X_2	Response	Rank	Vertex Number	Times Retained
Coordinates of	50.00	30.00	97.50	B	2	2
retained vertexes	10.00	70.00	97.00	N_a	3	2
Σ	60.00	100.00				
$\bar{P} = \Sigma/k$	30.00	50.00				
W	50.00	70.00	85.50	W_a	4	1
$(\bar{P} - W)$	-20.00	-20.00				
$R = \bar{P} + (\bar{P} - W)$	10.00	30.00		R	5	0

Figure 7.6 Calculations for generating vertex #5 in Figure 7.4. $\mathbf{N_a}$ is the absolute next-to-the-worst vertex; $\mathbf{W_a}$ is the absolute worst vertex. See text for details.

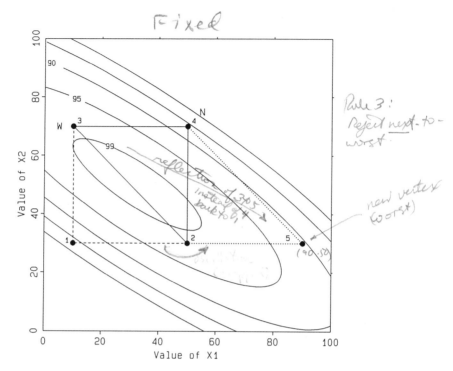

Fixed

Rule 3: Reject next-to-worst.

reflection

Instead of 3-0-5 back to 0, 1

new vertex (worst)

(90, 50)

Figure 7.7   Effect of application of Spendley, Hext, and Himsworth's Rule 3. **N** and **W** designate rows in the worksheet of Figure 7.8. See text for details.

nation, and the new vertex #5 will be the worst vertex in the new simplex. This will lead to oscillation between the two factor combinations given by vertexes #1 and #4, and the simplex will become stranded on the ridge. Although this is not a severe problem in Figure 7.4 (the oscillating simplexes straddle the optimum), oscillation about a ridge can be frustrating when it occurs far from the optimum.

To avoid this problem of oscillation, Spendley, Hext, and Himsworth [1] introduced an additional rule (their "Rule 3"): if the new vertex is the worst vertex, do not apply Rule 1; instead, reject the vertex with the next-worst response in the current simplex. The effect of this additional rule is shown in Figure 7.7.

In Figure 7.7 the vertexes in the current simplex are connected by solid lines (vertexes #2, #3, and #4). The new vertex (vertex #4) is the worst vertex in this current simplex. Thus, Rule 3 applies: instead of rejecting the worst vertex in the current simplex (vertex #4), the vertex with the next-worst response (vertex #3) will be rejected.

On a simplex worksheet, the vertex to be rejected is placed on the row labeled **W**. Thus, in Figure 7.7 the label **W** does not identify the absolute worst vertex, but instead identifies the vertex that has been placed in the wastebasket row, the row labeled **W** on the worksheet displayed in Figure 7.8. Similarly, the designation **N** does not identify the absolute next-to-the-worst vertex, but instead identifies the vertex that has been placed in the row labeled **N** on the worksheet in Figure 7.8.

Simplex No. _2_ → _3_	Factor		Response	Rank	Vertex Number	Times Retained
	X_1	X_2				
Coordinates of	50.00	30.00	97.50	B	2	2
retained vertexes	50.00	70.00	85.50	N	4	1
Σ	100.00	100.00				
$\bar{P} = \Sigma/k$	50.00	50.00				
W	10.00	70.00	97.00	W	3	2
$(\bar{P} - W)$	40.00	-20.00				
$R = \bar{P} + (\bar{P} - W)$	90.00	30.00	90.50	R	5	0

Figure 7.8 Calculations for generating vertex #5 in Figure 7.7. See text for details.

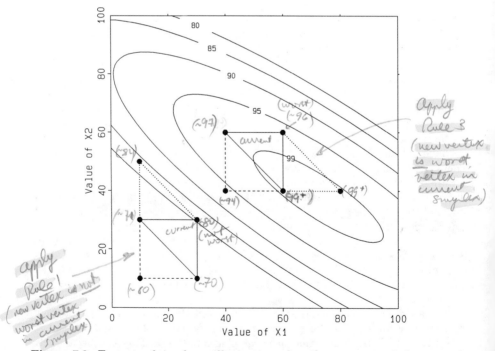

Figure 7.9 Two sets of simplexes illustrating when the new vertex is (center set) and is not (lower left set) the worst vertex in the current simplex. See text for details.

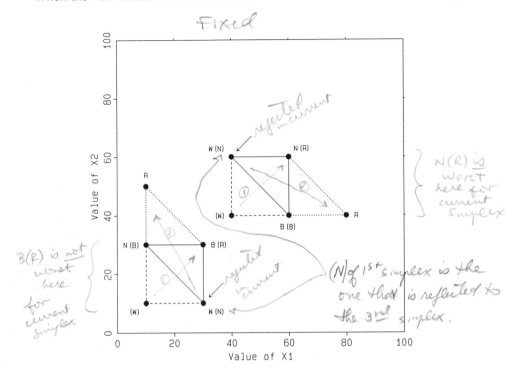

Handwritten annotations: Fixed. rejected in current. N(R) is worst here for current simplex. B(R) is not worst here for current simplex. (N) of 1st simplex is the one that is rejected to the 3rd simplex.

Figure 7.10 Designation of rows into which vertexes are placed for the vertexes shown in Figure 7.9. Letters in parentheses indicate rows in the previous simplex worksheet. Letters *not* in parentheses indicate rows in the current simplex worksheet. The arrows emphasize that in both cases, the vertex rejected in the current simplex is the vertex that was in the row labeled **N** in the previous simplex.

When the absolute next-to-the-worst vertex (labeled **W** in Figure 7.7) is rejected, a different vertex #5 results (see the calculations in Figure 7.8). Thus, Rule 3 avoids the problem of oscillation of the fixed-size simplex about a ridge. (This rule is also responsible, in part, for the final circling of the fixed-size simplexes in Figures 3.26 and 3.27.)

Figure 7.9 shows two sets of simplex moves. In each set, the current simplex vertexes are connected by solid lines. In the set shown at the lower left of the diagram, the new vertex is *not* the worst vertex in the current simplex. Thus, Rule 1 is the only rule that needs to be applied. However, in the set shown near the center of the diagram, the new vertex *is* the worst vertex in the current simplex. In this case, Rule 3 must be applied.

In Figure 7.10, the contours of constant response have been removed for clarity, and the vertexes shown in Figure 7.9 have been labeled. The letters beside each vertex indicate rows into which vertexes are placed for the various simplex moves. Letters in parentheses indicate rows in the previous simplex worksheet. Letters *not* in parentheses indicate rows in the current simplex worksheet.

Detailed study of Figure 7.10 reveals that in each of the two cases, the vertex rejected in the current simplex (the vertex in the row labeled **W**) is the vertex that

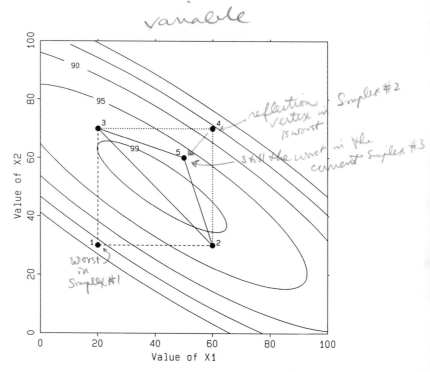

Figure 7.11 Results of a variable-size simplex iteration for which the new vertex (#5) is the worst vertex. See text for details.

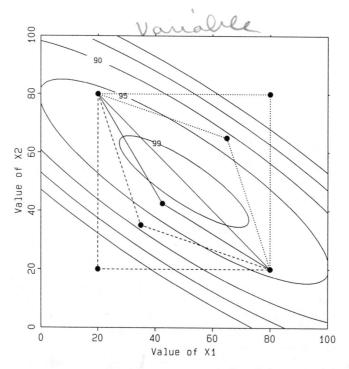

Figure 7.12 Partial oscillatory collapse of the variable-size simplex on a ridge. See text for details.

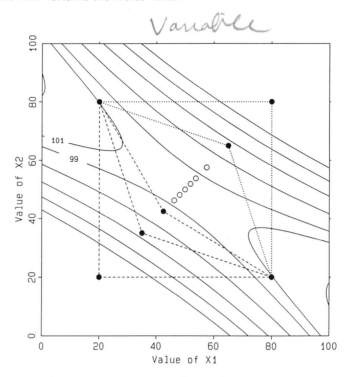

Figure 7.13 Catastrophic oscillatory collapse of the variable-size simplex on a saddle surface. See text for details.

was in the row labeled **N** in the previous simplex (see arrows). Consider the set of moves shown at the lower left of the figure: the vertex labeled **W(N)** is currently in the row labeled **W**, and was in the row labeled **N** in the previous simplex. Now consider the set of moves shown near the center of the figure: the vertex labeled **W(N)** is currently in the row labeled **W** (even though it is not the vertex with absolute worst response), and was in the row labeled **N** in the previous simplex. Thus,

> *whether or not the new vertex is the worst vertex in the current simplex, the vertex that will be rejected in the current simplex is the vertex that was in the row labeled **N** in the previous simplex.*

This observation makes it possible to combine Spendley, Hext, and Himsworth's Rules 1 and 3 [1] into a new rule, the rule we have used in presenting the fixed-size simplex algorithm (see Table 3.1): *always* transfer the current row labeled **N** to the row labeled **W** on the next worksheet.

Variable-Size Simplex

Figure 7.11 shows the results of a variable-size simplex iteration for which the new vertex is the worst vertex. The solid lines connect the vertexes in the new, current simplex (vertexes #2, #3, and #5). The dashed lines connect this current simplex with the vertex that was rejected in the previous simplex (vertex #1). The dotted

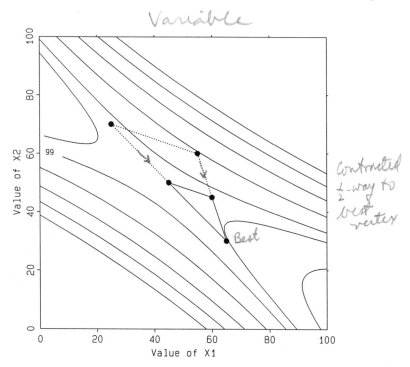

Figure 7.14 The "massive contraction" of Nelder and Mead [2]. When the new vertex is the worst vertex, all vertexes are contracted halfway toward the best vertex. This requires the evaluation of response at the k new vertexes.

lines connect this current simplex with the nonretained reflection (vertex #4). Because the simplex straddles the ridge, the new vertex (vertex #5) has the worst response in the new simplex.

   If each variable-size simplex move were always initiated with Spendley, Hext, and Himsworth's Rule 1 (that is, reject the vertex that has the worst response), the simplex might go into "oscillatory collapse." Partial oscillatory collapse is illustrated in Figure 7.12, where the first move results in a reflection followed by a C_W contraction. The second move is also a reflection followed by a C_W contraction, resulting in the current simplex (the three vertexes joined by solid lines). At this point a vertex other than the C_W contraction would be rejected, and the simplex would begin to contract about the optimum.

   If the response surface contains multiple optima or a local "dip" in response, then oscillatory collapse can be catastrophic and final. This is illustrated in Figure 7.13 for a saddle point. In the original simplex, two of the vertexes lie high on the response surface on either side of the saddle; the third vertex lies far down the hill. In this example, as the simplex goes into oscillatory collapse, the reflection and contraction vertexes come up onto the saddle but can never produce a response that is greater than the response at the other two vertexes. If only Rule 1 is used, the original two-dimensional simplex will eventually collapse into a one-dimensional line segment and become completely stranded.

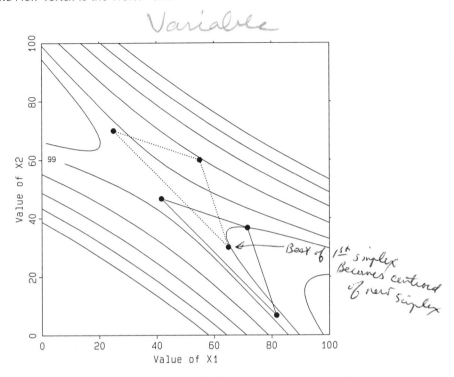

Figure 7.15 The translation move of Ernst [*3*]. When the new vertex is the worst vertex, all vertexes are shifted by an amount equal to the distance between the centroid and the vertex with best response. This requires the evaluation of response at the $k + 1$ new vertexes.

In the paper that introduced the variable-size simplex, Nelder and Mead [2] recognized the possibility of oscillatory collapse. They recommended that if the new vertex is the worst vertex (this will happen only after a C_R or C_W move), the simplex algorithm should be halted temporarily while a "massive contraction" is carried out. A massive contraction is shown in Figure 7.14. All old vertexes have been contracted halfway toward **B** along each line segment joining them with **B**. The response at each new vertex is evaluated, the vertexes are sorted from absolute best to absolute worst, and the algorithm is restarted. The effect of the massive contraction is to move the simplex away from troublesome regions of the response surface, to move the simplex on toward the optimum.

Although the massive contraction is straightforward, it is best suited to numerical optimizations for which the response can be calculated with minimum effort. This was the application for which the Nelder and Mead variable-size simplex [2] was originally intended–finding the optimum combination of parameter estimates to minimize the sum of squares of residuals when fitting nonlinear models to data (see Figure 10.2).

When the responses at each vertex must be evaluated experimentally, the massive contraction loses some of its appeal. A massive contraction requires the evaluating k new vertexes–that is, carrying out k new experiments.

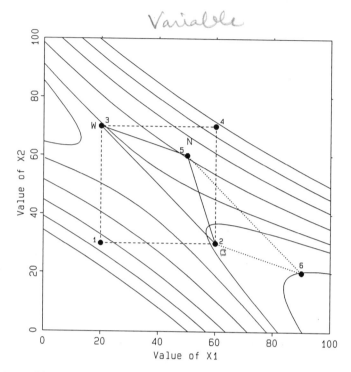

Variable

Figure 7.16 The move suggested by King [4]. When the new vertex is the worst vertex, reject the vertex that was on the row labeled **N** in the previous simplex. This requires the evaluation of response at only one new vertex.

Another disadvantage of the massive contraction is the abrupt decrease in the volume of factor space spanned by the simplex. In one dimension, a massive contraction decreases the one-dimensional volume (length) of the line segment by one-half. In two dimensions (shown in Figure 7.14), a massive contraction decreases the two-dimensional volume (area) of the triangle by one-fourth. In three dimensions, a massive contraction decreases the three-dimensional volume of the tetrahedron by one-eighth. And so on into higher dimensions. In the presence of noise, it is best to keep the variable-size simplex as large as possible for as long as possible. The massive contraction works against this, severely so in higher dimensional factor spaces.

Ernst [3], one of the first to use the variable-size simplex in an experimental environment, recognized the experimentalist's need for avoiding not only oscillatory collapse but also the abrupt decrease in volume that results from the massive contraction. As an alternative to the massive contraction, Ernst suggested that when the new vertex is the worst vertex, the simplex algorithm should be halted temporarily while a "translation" of the simplex is carried out. This translation is shown in Figure 7.15. The vector between the centroid and the best vertex in the current simplex is added to each of the vertexes (including **B**). The former best vertex **B** becomes the centroid of the new, translated simplex. Because all vertexes are moved the same distance and in the same direction, the resulting simplex has the same volume as the previous simplex.

Simplex No. 1 → 2	Factor					
	x_1	x_2	Response	Rank	Vertex Number	Times Retained
Coordinates of	60.00	30.00	101.50	B	2	1
retained vertexes	20.00	70.00	99.05	N	3	1
Σ	80.00	100.00				
$\bar{P} = \Sigma/k$	40.00	50.00				
W	20.00	30.00	87.00	W	1	1
$(\bar{P} - W)$	20.00	20.00				
$R = \bar{P} + (\bar{P} - W)$	60.00	70.00	82.00	R	4	0
$(\bar{P} - W)/2$	10.00	10.00				
$C_W = \bar{P} - (\bar{P} - W)/2$				C_W		0
$C_r = \bar{P} + (\bar{P} - W)/2$	50.00	60.00	89.90	C_r	5	0
$E = R + (\bar{P} - W)$				E		0

Figure 7.17 Calculations for generating vertex #5 in Figure 7.16. See text for details.

Although Ernst's modification avoids the problem of an abrupt decrease in volume of the simplex, it does not avoid the need for carrying out extra experiments. In fact, the translational move requires $k + 1$ new evaluations of response, even more discouraging for the experimentalist.

King [4] was the first to suggest an ingeniously efficient modification of the original variable-size algorithm to avoid oscillatory collapse. Rather than carry out a massive contraction or a translation, King suggested simply using Spendley, Hext, and Himsworth's Rule 3 [1]: if the new vertex is the worst vertex, reject the vertex with the next-worst response in the current simplex. That is, reject the vertex that was in the row labeled **N** in the previous simplex. In short, *always* transfer the current row labeled **N** to the row labeled **W** on the next worksheet.

The effect of this rule is shown in Figure 7.16. The vertexes in the current simplex are connected by solid lines (vertexes #2, #3, and #5). The dashed lines connect the current simplex with the previously rejected vertex (#1) and the failed reflection vertex (#4). The calculations giving rise to vertex #5 are shown in Figure 7.17. In this example, the new vertex #5 is the worst vertex. Figure 7.18 shows the calculations for the next simplex move. Note that the contents of the row labeled **N** in the old worksheet (Figure 7.17) have been transferred to the row labeled **W** in the new worksheet shown in Figure 7.18. The result is a reflection (vertex #6) that will lead to an attempted expansion.

The advantages of King's modification are clear:

Simplex No. 2 → 3

	X_1	X_2	Response	Rank	Vertex Number	Times Retained
Coordinates of	60.00	30.00	101.50	B	2	2
retained vertexes	50.00	60.00	89.90	N	5	1
Σ	110.00	90.00				
$\bar{P} = \Sigma/k$	55.00	45.00				
W	20.00	70.00	99.05	W	3	2
$(\bar{P} - W)$	35.00	-25.00				
$R = \bar{P} + (\bar{P} - W)$	90.00	20.00	105.02	R	6	0
$(\bar{P} - W)/2$						
$C_W = \bar{P} - (\bar{P} - W)/2$				C_W		0
$C_r = \bar{P} + (\bar{P} - W)/2$				C_r		0
$E = R + (\bar{P} - W)$				E		0

Figure 7.18 Calculations for generating vertex #6 in Figure 7.16. See text for details.

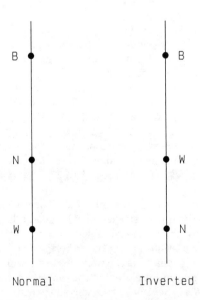

Figure 7.19 Normal (left) and inverted (right) rankings for **N** and **W**. The inverted rankings result when the new vertex is the absolute worst vertex. See text for details.

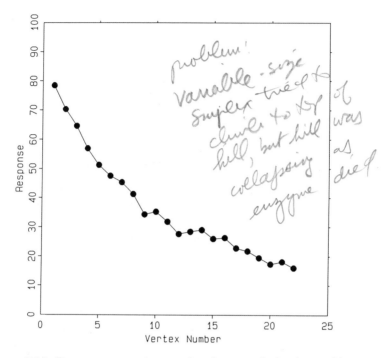

Figure 7.20 Response vs. vertex number for a maximization problem.

1. The simplex does not need to be halted temporarily.
2. It is not necessary to carry out k or $k + 1$ new evaluations of response. Only one new evaluation is required, independent of k.
3. The simplex does not undergo an abrupt decrease in volume.
4. No special rules must be created. Always transfer the current row labeled **N** to the row labeled **W** on the next worksheet.

It is curious that in the variable-size simplex, if the new vertex is the worst vertex, then on the current move the variable-size simplex rules logically exclude a C_R contraction. This can be understood by reference to Figure 7.19 and Table 4.1. The two number lines shown in Figure 7.19 represent response. It is assumed here that higher values of response are more desirable. The arrangement shown on the left in Figure 7.19 is the normal ranking of vertexes, in which the most desirable response is found on the row labeled **B**, the absolute next-to-the-worst response is found on the row labeled **N**, and the absolute worst response is found on the row labeled **W**. The arrangement shown on the right is the inverted ranking of the **N** and **W** vertexes that results when the new vertex is the worst vertex and the absolute next-to-the-worst vertex gets into the wastebasket vertex: the most desirable response is still found on the row labeled **B**, but the absolute next-to-the-worst response is found on the row labeled **W**, and the absolute worst response is found on the row labeled **N**.

Table 4.1 gives the rules for the variable-size simplex. After calculating and evaluating the reflection vertex **R**, the contraction rule 2.C is invoked if $\mathbf{R} < \mathbf{N}$. It

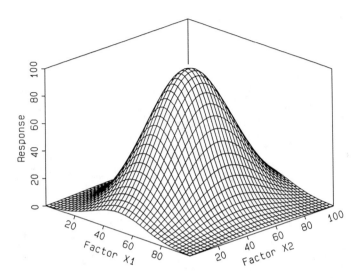

Figure 7.21   A two-factor response surface for an enzyme system at 1:00 p.m.

is clear from the right side of Figure 7.19 that if the response at the reflection vertex **R** is worse than the response on the row labeled **N**, the response at **R** must also be worse than the response on the row labeled **W**. Thus, because **R** < **N** and therefore **R** < **W**, the conditions of rule 2.C.i *cannot* be true: "If **R** ≥ **W**, calculate and evaluate C_R." Instead, the conditions of rule 2.C.ii *must* be true: "If **R** < **W**, calcu-

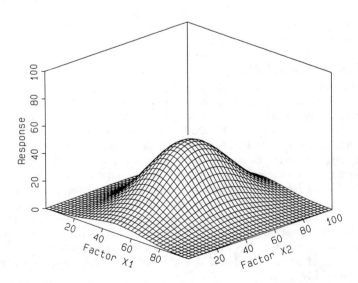

Figure 7.22   The two-factor response surface for an enzyme system at 3:00 p.m.

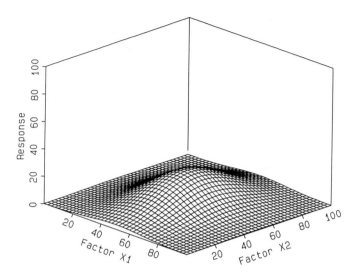

Figure 7.23 The two-factor response surface for an enzyme system at 5:00 p.m.

late and evaluate C_W." Thus, if the new vertex is the worst vertex and a contraction is to be carried out, then it must be a C_W contraction.

UNSTABLE SYSTEMS

Figure 7.20 shows the results of a variable-size simplex optimization carried out by an undergraduate student one afternoon. The intent was to maximize the rate of an enzyme-catalyzed reaction by varying pH (factor x_1) and concentration of enzymatic cofactor (factor x_2). It is clear that the reaction rate is not being maximized.

Careful scrutiny of the student's worksheets indicated that he was carrying out the simplex calculations correctly. The variable-size simplex expanded initially, and then began a series of C_W contractions. The only unusual features of the contracting, converging simplex were the very small number of C_R contractions and the large number of $k + 1$ rule violations. Reevaluation of response for the $k + 1$ rule violations gave consistently slower rates than when evaluated initially. From a statistical point of view, C_W and C_R contractions have the same probability of occurring, so the lack of C_R contractions was puzzling.

The student had "good hands" and performed the experiments as instructed. It was not possible to find fault with his experimental technique. However, inspection of the lab bench revealed the source of the problem: the enzyme, dissolved in 10 mL of a saline solution, was sitting out in the open.

Good laboratory practice for this enzyme requires that it be stored in solution in a refrigerator at about 4°C to keep it from thermally denaturing (i.e., to keep it from losing its catalytic activity). When enzyme is needed, the container is taken from the refrigerator, a small amount of solution is removed, and the container is replaced in the refrigerator. The student had not been informed

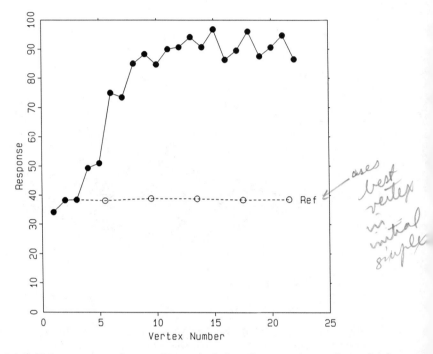

Figure 7.24 Reference experiments (open circles) at frequent intervals to check the system for drift.

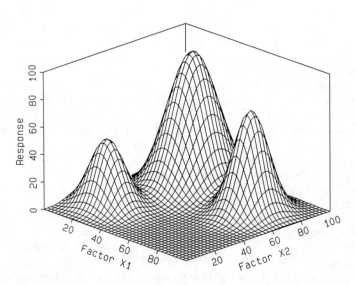

Figure 7.25 A response surface with multiple optima. Compare with the behavior shown in Figures 7.26 and 7.27.

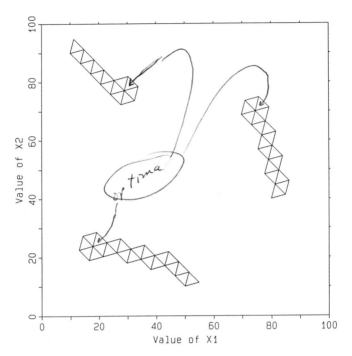

Figure 7.26 Behavior of fixed-size simplexes on a response surface with multiple optima.

of this experimental detail (the fault of SND, not the fault of the student), and had simply left the container of enzyme solution on the lab bench next to the other reagents. The results shown in Figure 7.20 could now be explained.

Figure 7.21 shows the hypothetical response surface for the fresh enzyme at 1:00 p.m. If the correct combination of pH and cofactor concentration (factors x_1 and x_2, scaled in Figure 7.21) is found, then maximum reaction rate (response) can be achieved.

Unfortunately, the enzyme was dying. Figure 7.22 shows the response surface for the enzyme at 3:00 p.m. after about half of the enzyme molecules had denatured. Figure 7.23 shows the response surface at 5:00 p.m. after about three-fourths of the enzyme molecules had denatured.

The variable-size simplex was trying to climb to the top of the hill, but the hill was collapsing. The large number of $k + 1$ rule violations is easy to understand because the best vertexes would have been early vertexes and would persist as the subsequent vertexes obtained poorer and poorer responses. The large number of C_W contractions is understandable when it is realized that a collapsing response surface will produce responses at the reflection vertex that will tend to be lower than the rejected **W** vertex that had been obtained earlier. All of this points out the need to work with a stable system. The behavior shown in Figure 7.20 could have been avoided if the enzyme had been kept in the refrigerator.

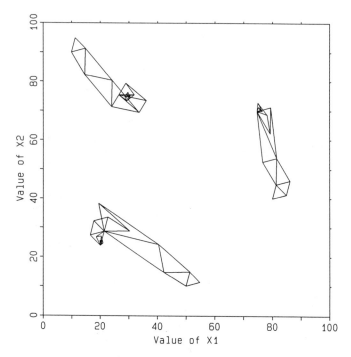

Figure 7.27 Behavior of variable-size simplexes on a response surface with multiple optima.

Figure 7.24 illustrates a method for monitoring the stability of a system. After a predetermined number of vertexes have been evaluated, a reference experiment is carried out (the open circle at vertex #6). This experiment typically corresponds to the experimental conditions associated with the best vertex in the initial simplex. If the reference conditions produce approximately the same result as was obtained earlier, then the simplex is allowed to proceed. After a certain number of subsequent vertexes have been evaluated, the reference experiment is carried out again. This is continued at regular intervals. In the example shown in Figure 7.24, the system appears to be stable—the reference experiment produces approximately the same result each time.

If the results of the reference experiments show an upward or downward trend in response, then the system is not stable. One possible action would be to stop the sequential simplex and take measures to bring the system into a state of statistical control.

If the system is changing because of proportional increases or decreases in the overall response surface as shown in Figures 7.21–7.23 (up-and-down motion), then it might be possible to model the change in response and "correct" the observed responses for this drift. However, if it is suspected that the system is changing because the response surface is moving across the factor space (sideways motion), then this strategy is probably inappropriate.

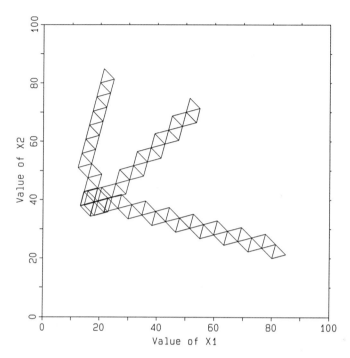

Figure 7.28 Behavior of fixed-size simplexes on a response surface with one optimum.

MULTIPLE OPTIMA

Investigators often worry that their systems have multiple optima. An example of a response surface possessing multiple optima is shown in Figure 7.25.

Systems with multiple optima do exist in the area of column separations [5]. Catalyst systems have the potential for multiple optima (e.g., a catalyst that is active in its monoligated or triligated form, but not in its diligated form will have two optimal regions of ligand concentration). Polymer systems might also exhibit multiple optima. Fortunately, the number of systems having multiple optima is small: there is no optimization strategy other than an exhaustive grid search that can guarantee that the global, or overall, optimum has been found.

If a system does possess multiple optima, then starting small simplexes from different regions of factor space might reveal some of these multiple optima. Figure 7.26 shows the progress of small fixed-size simplexes across the factor space of Figure 7.25. Figure 7.27 shows the progress of small variable-size simplexes. When several simplex optimizations of the same system reveal multiple optima, then it is possible to choose from among the optima that have been found. But there is no guarantee that the system does not still possess a better optimum that has not yet been discovered.

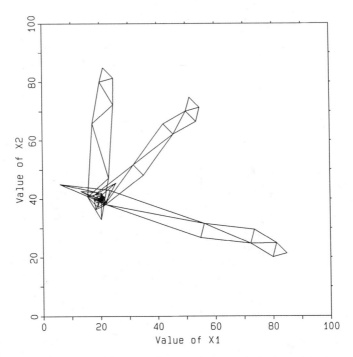

Figure 7.29 Behavior of variable-size simplexes on a response surface with one optimum.

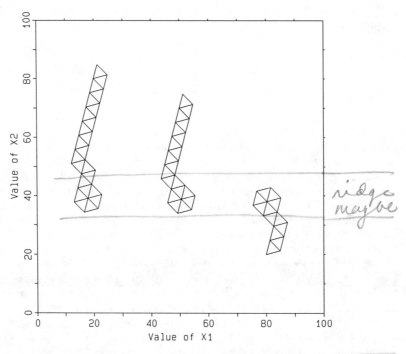

Figure 7.30 Behavior of fixed-size simplexes on a response surface with apparent multiple optima.

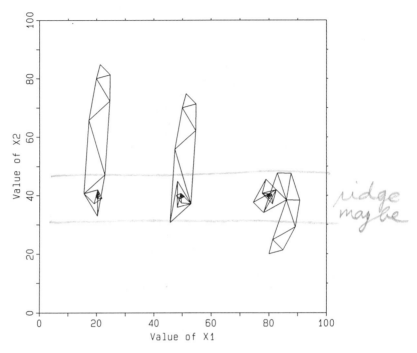

Figure 7.31 Behavior of variable-size simplexes on a response surface with apparent multiple optima.

If a response surface exhibits only one optimum, then multiple small simplexes started from different regions of factor space will behave as shown in Figure 7.28 for the fixed-size simplex and in Figure 7.29 for the variable-size simplex.

Occasionally, multiple small simplexes will behave as shown in Figure 7.30 and 7.31. When this happens, the response at these optima is usually about the same. What is also curious about this system is that the various optima all have different values associated with the factor x_1, but they all have about the same value of x_2.

This behavior might be attributed to a response surface that looks like that shown in Figure 7.32, but it can also happen (and probably happens more frequently) with response surfaces that look like that shown in Figure 7.33. The response surface shown in Figure 7.33 is a ridge. The factor x_2 is seen to have a strong influence on the response, but the response is almost totally insensitive to the factor x_1. Thus, the behavior shown in Figure 7.30 and 7.31 could arise from the this underlying ridge: the simplexes will adjust x_2 to its optimal value, but the value obtained for x_1 is totally random—any value of x_1 will give the same response for a given value of x_2.

SAFETY CONSIDERATIONS

Figure 7.34 shows a manufacturing facility before simplex optimization. Figure 7.35 shows the same facility after simplex optimization. Do not do this.

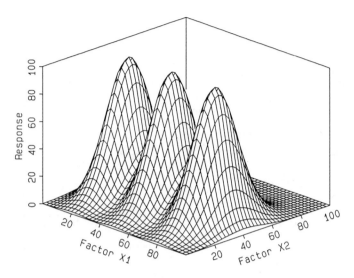

Figure 7.32 Response surface with multiple optima. Compare with the behavior shown in Figures 7.30 and 7.31.

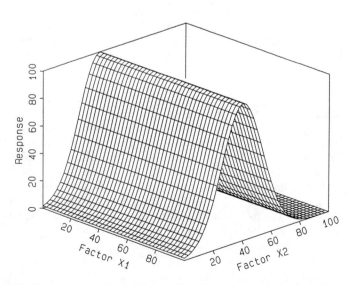

Figure 7.33 Response surface with a ridge. Compare with the behavior shown in Figures 7.30 and 7.31.

Figure 7.34 A manufacturing facility before simplex optimization.

A number of years ago we gave a sequential simplex optimization short course in Houston. One of the participants was a young engineer from a chemical manufacturing company in northeast Texas. The participant was enthusiastic about taking the simplex methods back to his company and improving many of the processes there. We taught him as much as we knew then, sent him on his way, and wished him well.

Two weeks later one of us opened the morning paper and saw on the front page a full-color, four-column photograph of a devastating explosion that had occurred at this chemical manufacturing company in northeast Texas. Stricken with fear that we might have been indirectly responsible for this disaster, we made a few surreptitious telephone calls. As it turned out, the explosion had nothing to do with the participant in our course. But it did make us stop and think: variable-size simplexes can be dangerous!

Consider a hypothetical situation. Figure 7.36 shows the bounded factor space for a two-factor chemical manufacturing facility. The elliptical contours of constant response suggest that there is an optimum at approximately $x_1 = 70$, $x_2 = 30$. The

Figure 7.35 A manufacturing facility after simplex optimization. **Do not do this!**

shaded region near the bottom right of the factor space indicates explosive combinations of x_1 and x_2. The white band adjacent to this shaded region is an area where the system will behave in an unusual fashion (e.g., it might produce a yellow-green gas that is not normally present), but no explosion will take place. Remember that the response surface, the explosive region, and the band of unusual behavior are not known initially, but must be discovered by experiment.

In Figure 7.36 the variable-size simplex begins at known operating conditions: $x_1 = 10$, $x_2 = 10$ with a moderate step size of 10 in each factor. The responses at the new vertexes in the initial simplex produce responses that are better than the facility has ever produced, so it is relatively easy to convince the plant operator and upper management that the simplex is a good experimental design. The safety officer, who has been monitoring the experiments, is pleased that nothing unusual has occurred during any of these experiments.

The first simplex move is a reflection that produces even better results. This leads to a retained expansion. Management is pleased. Perhaps tremendous profit

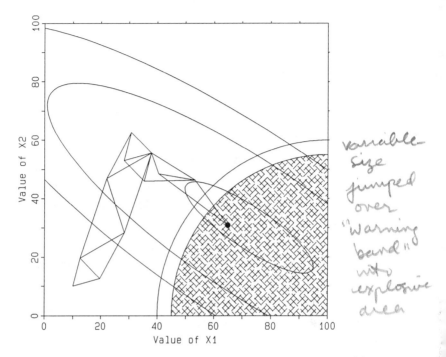

variable size jumped over "warning band" into explosive area

Figure 7.36 A variable-size simplex expanding into an explosive region of factor space (shaded area). See text for details.

can be realized if the response is increased further. This happens on the next reflection. The subsequent retained expansion produces even greater response. It looks as if the simplex can do no wrong. Quarterly profits are increasing.

The next reflection gives still greater response, and by now the whole plant is watching the simplex progress. Unfortunately, the expansion goes over the hill and is not kept, but the retained reflection pleases everyone. The next reflection results in a C_R contraction. This is followed by a C_W contraction, a simple reflection, and a subsequent reflection that results in a retained expansion. Finally, the simplex is producing improved responses again.

The next move results in a C_W contraction. This is followed by a reflection that produces a response that is better than the previous best. The expansion is attempted – and that is when the facility blows up.

Variable-size simplexes are dangerous. They can make relatively large leaps across the factor space. They can jump over regions that might otherwise give warning that unusual events are occurring and the safety officer should be consulted.

Consider an alternative hypothetical situation. Figure 7.37 again shows the bounded factor space for a two-factor chemical manufacturing facility. In Figure 7.37 the fixed-size simplex has a small step size of 5 in each factor. The fixed-size simplex moves doggedly toward the optimum, producing slightly better responses as it moves up the side of the hill, onto the ridge, and toward the optimum.

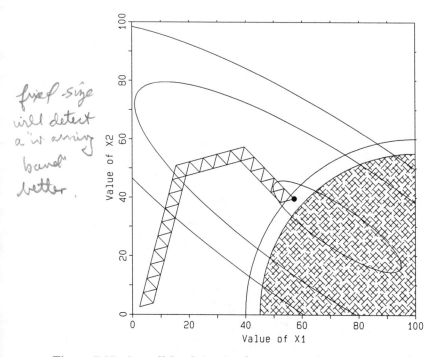

fixed-size
will detect
a "warning
band"
better.

Figure 7.37 A small fixed-size simplex encountering a warning region next to an explosive region of factor space (shaded area). See text for details.

In Figure 7.37, the last simplex is shown entering the band of unusual behavior. At this point, the safety officer becomes somewhat concerned because the reaction is producing a yellow-green gas that has never been seen coming from this reaction before. A wise safety officer will probably tell the experimenters to stop, to move back to conditions that are known to be safe until small-scale investigations can reveal the source of the yellow-green gas.

If systems do have warning regions that border catastrophic areas of factor space, there is a greater chance that these warning regions will be encountered if a small, fixed-size simplex is used. Variable-size simplexes are too risky to use in potentially hazardous environments—there is too great a chance that they will leap without warning into a catastrophic region.

Figure 7.34 shows a manufacturing facility before simplex optimization. Figure 7.35 shows the same facility after simplex optimization. Do not do this.

OTHER TYPES OF SIMPLEX ALGORITHMS

The most commonly used simplex algorithms are the fixed-size simplex and variable-size simplex discussed earlier. A number of modifications have been introduced in the literature, however.

Cave [6] investigated an alternative approach to dealing with boundary violations: placing the new vertex **R** at the boundary instead of assigning it an arbitrarily bad response. Several groups [7–9] developed weighted-centroid algorithms which place the centroid of the remaining hyperface closer to the better vertexes by weighting them more than the other vertexes. The simplex often becomes degenerate (loses one degree of freedom, or dimension, along which to search), so the weighted-centroid method gave rise to other further modifications such as the controlled weighted-centroid [10].

One modification to the simplex algorithm that has enjoyed a measure of success is the supermodified simplex [11]. The approach behind this algorithm is to not restrict the new vertex to just four possible points (**R**, **E**, **C**$_R$, and **C**$_W$) on a line connecting the wastebasket vertex **W** and the centroid $\bar{\mathbf{P}}$, but to allow the new vertex to be anywhere (with certain restrictions) along this line and its extensions, inside or outside the current simplex. In the supermodified simplex algorithm, a normal reflection vertex **R** is evaluated and, in addition, an experiment is carried out at the coordinates of the centroid, $\bar{\mathbf{P}}$; this now gives three data points on the **W**–$\bar{\mathbf{P}}$–**R** line. Three data points allow a second-order, or quadratic, model to be fit (see Chapter 9); the first derivative of this model can be used to predict where the best response should occur on the **W**–$\bar{\mathbf{P}}$–**R** line. An experiment is conducted at these conditions and the response compared with the response at **R**; whichever point gives the best response becomes the new vertex. The new point might be out beyond **R** (similar to an expansion vertex) or it might be within the current simplex (similar to a contraction vertex)—the difference is that the simplex does not have to expand to only twice its size or contract to only one-half its size. Restrictions are placed on the simplex: normally the new vertex cannot be further away from $\bar{\mathbf{P}}$ than 10 times the distance $\bar{\mathbf{P}}$–**W**, and cannot be closer to **P** than 0.1 times the distance $\bar{\mathbf{P}}$–**W**. Once the simplex has become small and the responses close, indicating it is in the region of the optimum, this latter restriction is removed to allow the simplex to contract around the optimum. If the quadratic model predicts the best response should occur beyond a boundary, the experimental conditions are set at the point where the **W**–$\bar{\mathbf{P}}$–**R** line intersects the boundary.

The supermodified simplex itself has come in for modifications. One of the drawbacks of the supermodified simplex is the need for an extra experiment (the one at $\bar{\mathbf{P}}$) for each simplex. A modification [12] used the average of the responses of all the retained vertexes to approximate the response at $\bar{\mathbf{P}}$, eliminating the need for an experiment at the centroid.

A fairly sophisticated simplex algorithm has been developed, called the composite-modified simplex [13, 14]. This algorithm incorporates a number of the other modifications, along with criteria for selecting when to invoke them. Among its features are polynomial fitting (similar to the supermodified simplex), fitting the new vertex to a boundary, "squashing" to a boundary (allowing the simplex to lose a degree of freedom and move along a boundary), and exploring a "suboptimal" direction (to avoid becoming stranded between two local optima).

All of these modifications to the simplex algorithm seem to offer some increase in the efficiency of the simplex (fewer experiments to reach the optimum and/or fewer convergences on false optima). However, all of these algorithms are much more complex than either the fixed-size or variable-size simplex algorithms. Some are difficult to carry out without the aid of a computer. Only the supermodified simplex has been used to any large extent outside the research group that developed it. Despite all these modified versions of the simplex, the original fixed-size simplex and variable-size algorithms are widely used; their simplicity and ease of operation remain very attractive.

REFERENCES

1. W. Spendley, G. R. Hext, and F. R. Himsworth, "Sequential application of simplex designs in optimisation and evolutionary operation," *Technometrics*, **4**, 441–461 (1962).
2. J. A. Nelder and R. Mead, "A simplex method for function minimization," *Comput. J.*, **7**, 308–313 (1965).
3. R. R. Ernst, "Measurement and control of magnetic field homogeneity," *Rev. Sci. Instrum.*, **39**, 998–1012 (1968).
4. P. G. King, "Automated development of analytical methods," Ph.D. Dissertation, Emory University, Atlanta, GA (1974).
5. J. L. Glajch and L. R. Snyder, *Computer-Assisted Method Development for High-Performance Liquid Chromatography*, Elsevier, Amsterdam, 1990.
6. M. R. Cave, "An improved simplex algorithm for dealing with boundary conditions," *Anal. Chim. Acta*, **181**, 107–116 (1986).
7. T. Umeda and A. Ichikawa, "A modified complex method for optimization," *Ind. Eng. Chem. Process Des. Dev.*, **10**(2), 229–236 (1971).
8. P. B. Ryan, R. L. Barr, and H. D. Todd, "Simplex techniques for nonlinear optimization," *Anal. Chem.*, **52**, 1460–1467 (1980).
9. E. R. Aberg and A. G. T. Gustavsson, "Design and evaluation of modified simplex methods," *Anal. Chim. Acta*, **144**, 39–53 (1982).
10. P. F. A. Van der Wiel, R. Maassen, and G. Kateman, "The symmetry-controlled simplex optimization procedure," *Anal. Chim. Acta*, **153**, 83–92 (1983).
11. M. W. Routh, P. A. Swartz, and M. B. Denton, "Performance of the super modified simplex," *Anal. Chem.*, **49**(9), 1422–1428 (1977).
12. L. R. Parker, Jr., M. R. Cave, and R. M. Barnes, "Comparison of simplex algorithms," *Anal. Chim. Acta*, **175**, 231–237 (1985).
13. D. Betteridge, A. P. Wade, and A. G. Howard, "Reflections on the modified simplex–I," *Talanta*, **32**, 709–722 (1985).
14. D. Betteridge, A. P. Wade, and A. G. Howard, "Reflections on the modified simplex–II," *Talanta*, **32**, 723–734 (1985).

Chapter 8

Desirability Functions

MULTIPLE CRITERIA OPTIMIZATION

In many experimental optimization problems, it is unusual to find only one response that needs to be optimized. Instead, there are usually several responses that must be considered. Additionally, many of the responses must be expressed as intensive properties (i.e., they should not depend on the size or throughput of the system [1]) and must be normalized by one or more factors and/or responses of the system [2]. In chemical engineering, for example, if x_{in} is the amount of a specific material that goes into a process and y_{out} is the amount of that same material that has not reacted, then *conversion* is expressed as $(x_{in} - y_{out})/x_{in}$. Similarly, cost is often usefully measured as dollars *per pound* of desired product actually produced, $y_\$/y_{lb}$.

In all of these cases, various ratios, penalties, and desirabilities can be used to specify quantitative objective functions [3].

OBJECTIVE FUNCTIONS

As stated by Beveridge and Schechter [2], "The aim of optimization is the selection, out of the multiplicity of potential solutions, of that solution which is the best with respect to some well-defined criterion. The choice of this criterion, the objective, is therefore an essential step in any study. . . . In general, economic criteria should be used, although technical forms are common." Today, quality characteristics are often an important contribution to the overall objective (Chapter 1).

An objective function is a mathematical relationship expressing the objective in terms of system factors and/or responses (Figure 8.1). Objective functions based on overall economic strategies tend to be highly complex. Objective functions based on more restricted technical and quality considerations are usually simpler.

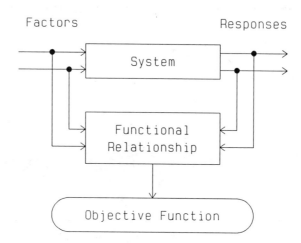

Figure 8.1 An objective function is a mathematical relationship expressing the objective in terms of system factors and/or responses.

Consider the system shown in Figure 8.2. It has two factors: x_1 = sugar, x_2 = fruit. It has two responses: y_1 = % alcohol, y_2 = flavor. The biochemical transform is apparently being utilized to make wine. What is the objective of the winemaking process?

If we assume that this is a commercial wine-making process, then the objective is probably to maximize the absolute rate of return on investment, $R = P/I$ where I is the investment and P is the profit. If we further assume that the wine-making process already exists, then the investment is mostly fixed and the objective now becomes the maximization of profit. Because the objective is now to maximize profit, what technical form will this objective take?

If we intend to sell this wine to a less discriminating clientele, we might want to make the alcohol content as high as legally possible and ignore the flavor of the wine. However, if we wish to sell this product to oenophiles of greater discrimination, we might want to put our effort into improving the flavor and ignore the alcohol content of the wine. The astute marketing manager will recognize immediately that what we really want to do in this situation is to improve both the alcohol content and the flavor—in that way we could sell to both markets to maximize our sales and, presumably, our profits.

An objective function could be used to formally indicate just how alcohol content and flavor should be combined into a single figure-of-merit to be optimized. For

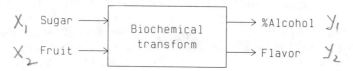

Figure 8.2 A systems theory view of a wine-making process.

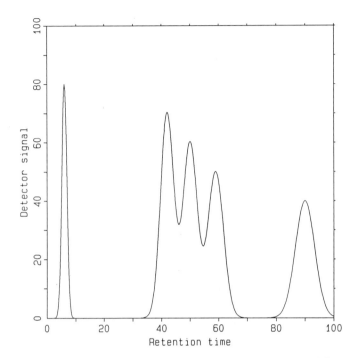

Figure 8.3 An incomplete liquid chromatographic separation of five inorganic ions. (After Smits, Vanroelen, and Massart [4].)

this example, the objective function might be simply the sum of the percent alcohol and the flavor as judged by a taste panel. Or if we want to emphasize the alcohol content, we might weight alcohol content twice as much as flavor. There might also exist target values of either or both responses; minimizing deviation from these target values might be the objective.

Considerations such as this illustrate an irony about the objective function: it is a highly *subjective* function. To write a proper objective function for this wine-making example, it would be helpful to have at hand the objective results of a market survey that measures the relative desirabilities of both percent alcohol and flavor for the intended consumers.

OBJECTIVE FUNCTIONS BASED ON RATIOS

Ratios are often used to construct objective functions. One economic ratio that is commonly used is the return on investment, $R = P/I$, discussed earlier. Another is the intensive profit-per-pound-produced, $y_\$/y_{lb}$. Although ratios are simple and attractive, they should be used with caution. Ratios can lead to unexpected results. This is well illustrated in the valuable paper by Smits, Vanroelen, and Massart [4]; the following discussion is based on their paper.

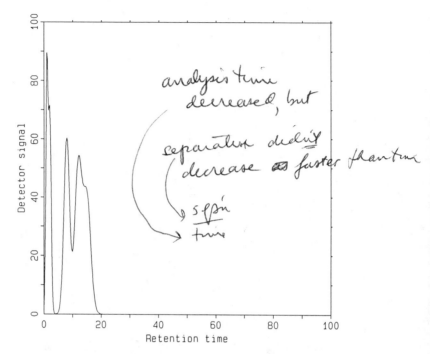

Figure 8.4 Results of optimization driven by the maximization of (separation/analysis time). (After Smits, Vanroelen, and Massart [4].)

Figure 8.3 shows an incomplete liquid chromatographic separation of five inorganic ions. The vertical axis represents detector signal, which is proportional to the concentration of the ions in the eluent. The horizontal axis represents the time after injection of the sample onto the chromatographic column. The separation can be measured and quantified: large measured values correspond to more complete separation; small measured values correspond to less complete separation.

Because this separation was to be used in an industrial environment, the time required to elute the last ion is also important: longer analysis times mean fewer samples per day; shorter analysis times will increase the daily sample throughput, clearly an economic advantage.

The ratio of (separation/analysis time) was chosen for maximization as the objective function. This seems reasonable. As the separation becomes more complete (desirable), the quantitative measure in the numerator will get larger and the objective function will get larger. As the analysis time decreases (desirable), the denominator will get smaller and again the objective function will get larger. Thus, maximizing the ratio of (separation/analysis time) should lead to improved separations and shorter analysis times.

Figure 8.4 shows a separation that results from the use of this optimization criterion. The objective function has been increased, but the separation of ions is worse now than when the optimization began. What went wrong?

What went wrong is that the denominator got smaller faster than the numerator got smaller. That is, the analysis time decreased faster than the separation degraded.

While the separation was going from bad to worse, the analysis time was going from good to better at a faster rate. The net result was a very fast "separation" that was almost totally worthless, even though the objective function ratio of (separation/analysis time) kept getting larger.

Objective functions based on ratios must be used with caution. An alternative is to avoid ratios by basing the optimization on only one of the components (e.g., separation) and establishing a threshold and penalty function for the other component (e.g., analysis time). Another alternative is to combine multiple responses into a single measure of performance that expresses the desirability of each combination.

OBJECTIVE FUNCTIONS BASED ON PENALTY FUNCTIONS

In the previous example involving maximization of separation and minimization of analysis time, practical considerations of sample throughput (e.g., analyses per day) often dictate a maximum permissible analysis time. If an analytical laboratory must carry out 15 analyses in an 8-hour day, then simple calculation suggests a maximum analysis time of approximately 30 minutes. This 30-minute maximum analysis time can be considered to be a threshold value: an analysis time less than 30 minutes might be desirable but wouldn't be especially beneficial, whereas an analysis time greater than 30 minutes would be undesirable, perhaps critically undesirable. Thus, an analysis time less than 30 minutes might not figure into any objective function calculations, but analysis times greater than 30 minutes should be taken into account. The objective function should be penalized if the analysis time exceeds 30 minutes.

Mathematically, penalty functions can be expressed as

$$p = p(y_j) \quad p(y_j) = 0 \text{ for } y_j \leq y_{jt} \tag{8.1}$$

$$p(y_j) = q(y_j - y_{jt}) \text{ for } y_j > y_{jt}$$

where p is the value of the penalty and y_{jt} represents the threshold value associated with y_j. The nature of $q(y_j - y_{jt})$ is subjective (of course) but usually follows one of three well-defined forms illustrated in Figure 8.5.

The first type of penalty function is an "infinite wall" illustrated at the top of Figure 8.5: $q(y_1 - y_{1t}) = -\infty$. Thus, violations of the threshold are considered to be infinitely bad. This type of penalty function is usually used for critical responses (those involving safety, for example).

A second type of penalty function is illustrated in the middle of Figure 8.5: $q(y_1 - y_{1t}) = b_1(y_1 - y_{1t})$ where b_1 is a slope or proportionality constant expressing the severity of the penalty ($b_1 = -\infty$ is equivalent to the "infinite wall"; $b_1 = 0$ is equivalent to no penalty). As the response gets farther away from the threshold value, the penalty becomes proportionally more severe. Again, the choice of b_1 is often subjective. (In process control, this type of penalty function is often used in one of the modes of control—e.g., the "P," or "proportional," in PID controllers for feedback loops.)

The third type of penalty function is illustrated at the bottom of Figure 8.5: $q(y_1 - y_{1t}) = b_2(y_1 - y_{1t})^n$ where n is usually 2 or greater. This is a power function

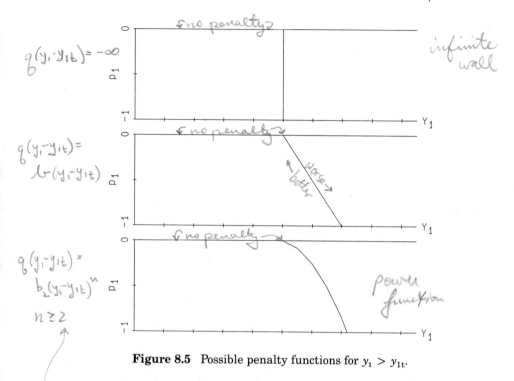

$$g(y_1 - y_{1t}) = -\infty$$

$$g(y_1 - y_{1t}) = b(y_1 - y_{1t})$$

$$g(y_1 - y_{1t}) = b_2(y_1 - y_{1t})^n$$

$$n \geq 2$$

Figure 8.5 Possible penalty functions for $y_1 > y_{1t}$.

that expresses the idea that large violations of the threshold value are much more serious than small violations, This is probably the most generally useful type of penalty function and is widely used in many areas. (Least-squares procedures for fitting models to data are based on this penalty function where $n = 2$, $b_2 = 1$, y_1 is the actual data point, and y_{1t} is the value estimated by the model.)

An example of the use of a penalty function applied to chromatographic separations may be found in the paper by Morgan and Deming [5].

DESIRABILITY FUNCTIONS

In the wine-making example shown in Figure 8.2, it was suggested that an objective function could be used that might add together the percentage alcohol and the flavor as judged by a taste panel. But how do you add the "apples" of percentage alcohol to the "oranges" of flavor rating? Harrington [6] states the problem well:

> In nearly all situations requiring human judgement, one is faced with a multiplicity of measures which must be balanced one against the other, weighted in accordance with their relative importance, compromised where these measures are mutually opposing, and variously manipulated to achieve an optimum judgement....
>
> If by some means the several properties could be measured in consistent units, or, even better, could be expressed as numbers on a dimensionless scale, then the arithmetic operations intended to combine these measures becomes feasible.

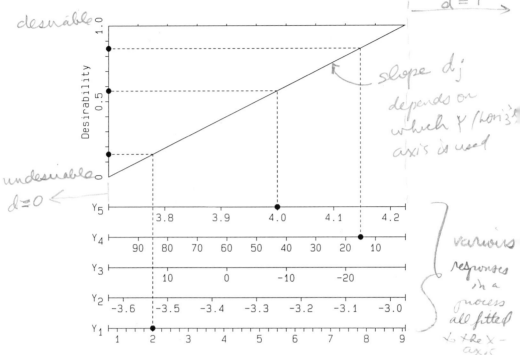

(handwritten annotations around figure:) desirable — undesirable — $d=0$ — $d=1$ — slope d_j depends on which Y (horiz?) axis is used — various responses in a process all fitted to the X-axis

Figure 8.6 Desirability as a first-order function of response. Undesirable responses at the left; desirable responses at the right. (After Lowe [*12*].)

Although Harrington proposed two specific forms for the "desirability function," the concepts are general and can be merged with concepts from the field of fuzzy logic [*7–11*] to yield useful objective functions for optimization.

Lowe [*12*] proposed a simple procedure for forming desirabilities from multiple responses. If y_{ju} and y_{jd} are measures of the most undesirable and most desirable values, respectively, of a response y_j, and if it is assumed that the desirability increases linearly in going from y_{ju} to y_{jd}, then the desirability contributed by this response is calculated as

$$d_j = 0 \qquad\qquad \text{for } y_j < y_{ju}$$

$$d_j = 1 \qquad\qquad \text{for } y_j > y_{jd}$$

$$d_j = (y_j - y_{ju})/(y_{jd} - y_{ju}) \qquad \text{for } y_{ju} \le y_j \le y_{jd} \qquad\qquad \textbf{(8.2)}$$

where "<" and ">" are again to be read as "worse than" and "better than" (Chapter 4). Note that d_j is unitless and ranges from 0 to 1.

The concept is illustrated in Figure 8.6. Along the left side at the top of the figure is a *desirability* axis ranging from 0 (undesirable) to 1 (desirable). Along the bottom of the figure are drawn five *response* axes, y_1–y_5. The response axes have been suppressed and expanded such that their most undesirable values are aligned

desirable

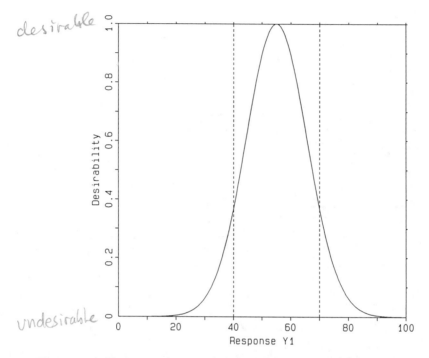

undesirable

Figure 8.7 Harrington's two-sided desirability function for $n = 2$. (After Harrington [6].)

vertically with the left side of the figure and their most desirable values are aligned vertically with the right side of the figure.

Running diagonally across Figure 8.6 from left to right is a transformation line that maps values of response onto values of desirability. This line is used by reading upward from a given value of response and leftward to the corresponding values of desirability. For example, a response value of $y_5 = 4.0$ corresponds to a desirability value of $d_5 = 0.57$. Similarly, $y_4 = 15$ becomes $d_4 = 0.85$, and $y_1 = 2$ becomes $d_1 = 0.15$. These results obtained graphically are identical to those obtained using Equation 8.2:

$$d_5 = (4.00 - 3.70)/(4.23 - 3.70) = 0.57$$

$$d_4 = (15 - 100)/(0 - 100) = 0.85$$

$$d_1 = (2.0 - 0.75)/(9.08 - 0.75) = 0.15 \tag{8.3}$$

Responses that lie to the right of the response ranges shown in Figure 8.6 would be assigned desirabilities of 1.00; response to the left of the figure would be assigned desirabilities of 0.00.

Harrington's desirability functions [6] do not assume a linear (first-order) relationship between response and desirability. The two-sided version is given by

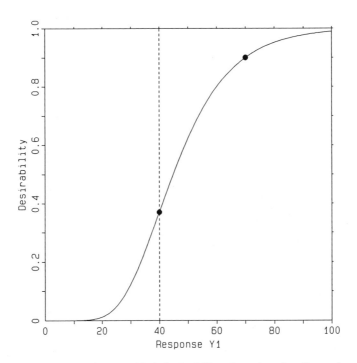

Figure 8.8 Harrington's one-sided desirability function for the ordered pairs (40.0, 0.37) and (70.0, 0.90). (After Harrington [6].)

$$d_j = \exp[-(|y_j'|)^n]\qquad(8.4)$$

where exp is the exponentiation function, n is a positive number ($0 < n < \infty$, not necessarily integral), y_j' is a linear transform of the response variable, y_j, such that $y_j' = -1$ when y_j is equal to the lower specification limit, y_{j-}, and $y_j' = +1$ when y_j is equal to the upper specification limit, y_{j+}, and $|y_j'|$ is the absolute value of y_j'. The use of upper and lower specification limits comes from concerns about product quality (Chapter 1). Any particular value of response, y_j, may be transformed into the corresponding y_j' by the relationship

$$y_j' = [y_j - (y_{j+} + y_{j-})/2]/[(y_{j+} - y_{j-})/2]$$
$$= [2y_j - (y_{j+} + y_{j-})]/(y_{j+} - y_{j-})\qquad(8.5)$$

which measures the distance of y_j from the midpoint between the upper and lower specification limits, $[(y_{j+} + y_{j-})/2]$, in units equal to one-half the spread between the upper and lower specification limits, $[(y_{j+} - y_{j-})/2]$. Figure 8.7 illustrates this two-sided desirability function for $n = 2$.

For one-sided specification limits a special form of the Gompertz growth curve is used:

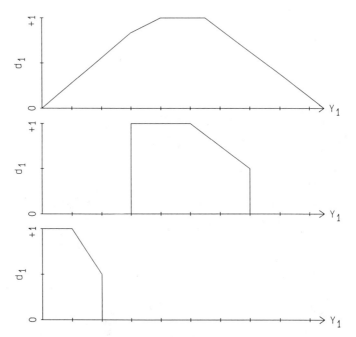

Figure 8.9 Free-form desirability functions constructed from straight-line segments.

$$d_j = \exp\{-[\exp(-y_j')]\} \qquad (8.6)$$

where $y_j' = 0$ at the single specification limit. The mapping of y_j onto y_j' is accomplished by choosing two ordered pairs of (y_j, d_j) and calculating $y_j' = -\ln[-\ln(d_j)]$ where ln is the natural logarithm function. From the resulting ordered pairs of (y_j, y_j'), the straight-line equation

$$y_j' = b_0 + b_1 y_j \qquad (8.7)$$

can be obtained, where b_0 is the intercept and b_1 is the slope. Figure 8.8 illustrates this one-sided desirability function for the ordered pairs (40.0, 0.37) and (70.0, 0.90).

These desirability functions are well suited to quality control work, but many alternative forms are possible. Some of the most useful versions of desirability functions are "free-form" graphic versions like those shown in Figures 8.9 and 8.10. These can often be developed in discussions between producers and consumers and can help lead to more meaningful product specification levels. Derringer and Suich [13] give examples. Other functions that are useful for representing desirability include the gaussian (bell-shaped) curve and the S-shaped logistic function.

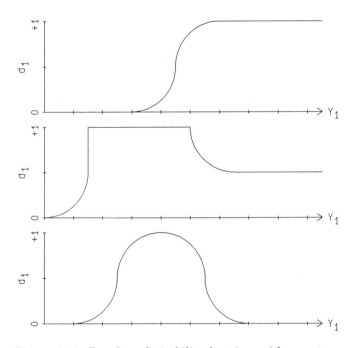

Figure 8.10 Free-form desirability functions with curvature.

OVERALL DESIRABILITIES

There are many ways the individual desirabilities d_1–d_n can be combined. A simple arithmetic average is one example. However, in any realistic situation, a [6]

> basic premise is this – if any one property is so poor that the product is not suitable to the application, that product will not be acceptable, *regardless of the remaining properties*. A structural plastic, for example, which is ideal in every way except that it would soften in the summer sun, would be completely useless in any location experiencing this degree of heat. It is also true that customer reaction to a product is based very largely on the less desirable properties of that product because these are the focus of potential trouble.

As Harrington goes on to point out [6], the mathematical model analogous to these psychological reactions is the geometric mean of the component d's, or

$$D = (d_1 d_2 \ldots d_n)^{1/n} \tag{8.8}$$

It is clear that if *any* d_j is zero, the associated D will also be zero. Furthermore, D is strongly weighted by the smaller d_js.

Figure 8.11 shows how D varies as a function of two d_js. The nth root (square root) relationship is clear in this representation. Note again that if either d_1 or d_2 goes to zero, D is zero regardless of the value of the other d.

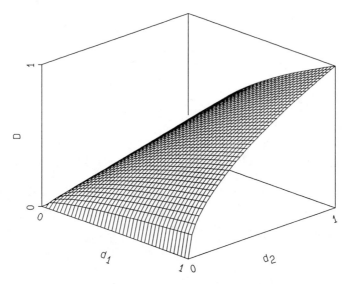

Figure 8.11 Illustration of how D (the overall desirability) varies as a function of two d_js according to Equation 8.8.

Figure 8.12 shows individual desirabilities, d_1 and d_2, as functions of two responses, y_1 and y_2. Mapping these desirabilities through Equation 8.8 gives Figure 8.13, which shows how the overall desirability D is affected by the individual *responses*, y_1 and y_2. Figures 8.14 and 8.15 suggest that more complicated mappings of responses onto desirabilities give rise to more complicated desirability surfaces that might contain multiple optima.

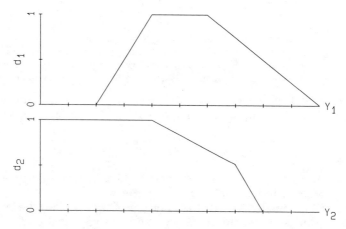

Figure 8.12 Individual desirabilities, d_1 and d_2, as functions of two responses, y_1 and y_2.

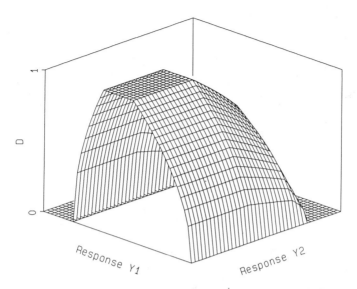

Figure 8.13 Overall desirability, D, plotted as a function of the individual responses, y_1 and y_2, mapped through Equation 8.8 using the individual desirabilities, d_1 and d_2, shown in Figure 8.12.

GENERAL COMMENTS

The ultimate mapping [6] would be to show D as a function of the system factors, but to do so presumes a knowledge of the relationships between each y_j and all x_is. However, because these relationships are not usually known at the beginning of a

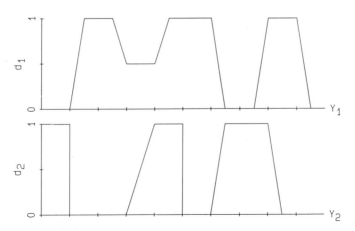

Figure 8.14 Polymodal individual desirabilities, d_1 and d_2, as functions of two responses, y_1 and y_2.

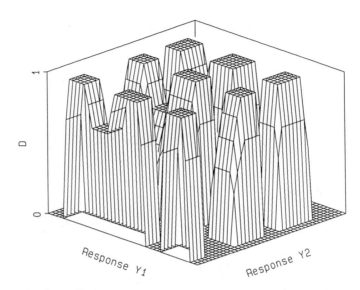

Figure 8.15 Overall desirability, D, plotted as a function of the individual responses, y_1 and y_2, mapped through Equation 8.8 using the individual desirabilities, d_1 and d_2, shown in Figure 8.14.

research project (after all, if they were known there would be no need for experimental optimization techniques such as the sequential simplex), such mappings are not usually possible.

Desirability functions have been used before in analytical chemistry to improve the quality of separations. The work of Laub and Purnell [14], Glajch and Kirkland [15], Sachok et al. [16], Morgan and Jacques [17], and Glajch and Snyder [18] may be consulted for examples.

We and others have found desirability functions to be especially useful for simplex optimization. However, care should be taken that the functions are well understood before the optimization is begun [4].

REFERENCES

1. S. N. Deming and S. L. Morgan, *Experimental Design: A Chemometric Approach*, Elsevier, Amsterdam, 1987.
2. G. S. G. Beveridge and R. S. Schechter, *Optimization: Theory and Practice*, McGraw-Hill, New York, 1970.
3. R. E. Steuer, *Multiple Criteria Optimization: Theory, Computation, and Application*, Wiley, New York, 1986.
4. R. Smits, C. Vanroelen and D. L. Massart, "The optimisation of information obtained by multicomponent chromatographic separation using the simplex technique," *Z. Anal. Chem.*, **273**, 1–5 (1975).

5. S. L. Morgan and S. N. Deming, "Optimization strategies for the development of gas-liquid chromatographic methods," *J. Chromatogr.*, **112**, 267–285 (1975).
6. E. C. Harrington, Jr., "The desirability function," *Indust. Qual. Control*, **21**, 494–498 (1965).
7. L. A. Zadeh, "Fuzzy sets," *Inform. Control*, **8**, 338–353 (1965).
8. L. A. Zadeh, "Yes, no, and relatively," *Chemtech*, **17**, 340–344 (1987).
9. A. Kandel, *Fuzzy Mathematical Techniques with Applications*, Addison-Wesley, Reading, MA, 1986.
10. A. Kaufmann and M. M. Gupta, *Introduction to Fuzzy Arithmetic: Theory and Applications*, Van Nostrand Reinhold, New York, 1985.
11. C. V. Negoita, *Expert Systems and Fuzzy Systems*, Benjamin/Cummings, Menlo Park, CA, 1985.
12. C. W. Lowe, "A report on a simplex evolutionary operation for multiple responses," *Trans. Inst. Chem. Eng.*, **45**, T3–T7 (1967).
13. G. Derringer and R. Suich, "Simultaneous optimization of several response variables," *J. Qual. Technol.*, **12**, 214–219 (1980).
14. R. J. Laub and J. H. Purnell, "Criteria for the use of mixed solvents in gas-liquid chromatography," *J. Chromatogr.*, **112**, 71–79 (1975).
15. J. L. Glajch, J. J. Kirkland, K. M. Squire, and J. M. Minor, "Optimization of solvent strength and selectivity for reversed-phase liquid chromatography using an interactive mixture-design statistical technique," *J. Chromatogr.*, **199**, 57–79 (1980).
16. B. Sachok, J. J. Stranahan, and S. N. Deming, "Two-factor minimum alpha plots for the liquid chromatographic separation of 2,6-disubstituted anilines," *Anal. Chem.*, **53**, 70–74 (1981).
17. S. L. Morgan and C. A. Jacques, "Response surface evaluation and optimization in gas chromatography," *J. Chromatogr. Sci.*, **16**, 500–505 (1978).
18. J. L. Glajch and L. R. Snyder, Eds., *Computer-Assisted Method Development for High-Performance Liquid Chromatography*, Elsevier, Amsterdam, 1990.

Chapter 9

Experimental Design

MODELS

After the sequential simplex has adjusted the factor levels to the region of the optimum, it is usually desirable to understand (in an operational way) the response in that region. That is, it is useful to know how the response y_1 varies with all of the x_k in the region of the optimum, to know the approximate relationship between y_1 and x_k. Sometimes this knowledge can come from data that have been acquired as the simplex converged into the region of the optimum (Chapter 5). Usually, however, it will come from a separate set of experiments specifically designed to acquire sufficient data to fit a mathematically linear model of suitable complexity. (A linear model is constructed of additive terms, each of which contains one, and only one, multiplicative parameter.)

There is an intimate relationship between experimental designs and mathematical models: experimental designs are usually constructed with specific models in mind. Thus, before we discuss experimental designs themselves, we will consider different mathematical forms that can be used as models to approximate the behavior of real systems [1].

Single-Factor Systems

Figure 9.1 illustrates the familiar straight-line relationship showing the response y_1 as a function of a single factor x_1. This model is often written

$$y = a + bx \tag{9.1}$$

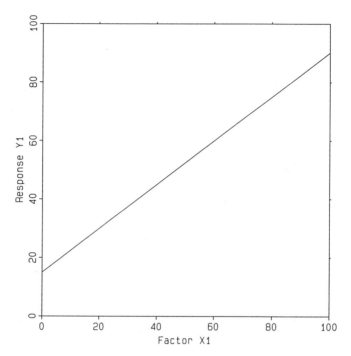

Figure 9.1 Graph of the straight-line model $\hat{y}_1 = \beta_0 + \beta_1 x_1$.

or

$$y = mx + b \qquad \text{(9.2)}$$

We will use y_1 for y to indicate the first type of response, tacitly acknowledging the existence of other types of responses $y_2, y_3, \ldots, y_j$. Similarly, we will use x_1 for x to indicate the first type of factor, again implicitly acknowledging the existence of other types of factors $x_2, x_3, \ldots, x_k$. We will use β_0 to indicate the intercept parameter of a model (a in Equation 9.1 and b in Equation 9.2) and β_1 to indicate the slope with respect to the factor x_1 (b in Equation 9.1 and m in Equation 9.2). Thus, Equations 9.1 and 9.2 will be written

$$y_1 = \beta_0 + \beta_1 x_1 \qquad \text{(9.3)}$$

If a model is to be used to describe the results of a set of n experiments, it can also be used to describe any individual experiment i in the set. Thus, Equation 9.3 is often written

$$y_{1i} = \beta_0 + \beta_1 x_{1i} \qquad \text{(9.4)}$$

where i can take on the values 1 to n. Further, because the model will probably not fit the data perfectly, some allowance must be made for discrepancies between the

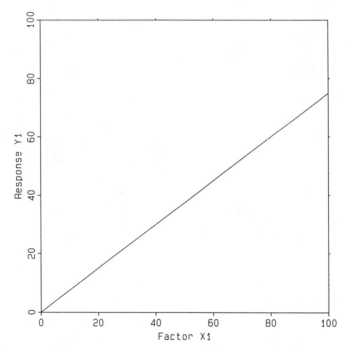

Figure 9.2 Graph of the straight line constrained to go through the origin, $\hat{y}_1 = \beta_1 x_1$.

response predicted by the model ($\hat{y}_{1i}$) and the response that is actually observed (y_{1i}). This discrepancy is usually called a *deviation* or a *residual* or an *error*, and is given the symbol r_{1i} or ε_{1i} ($= \hat{y}_{1i} - y_{1i}$). We will use the symbol r_{1i} to represent the residual. Thus, the complete probabilistic description of a straight-line model is

$$y_{1i} = \beta_0 + \beta_1 x_{1i} + r_{1i} \tag{9.5}$$

After the model has been fitted to the data, the deterministic part of it is written without the subscript i, without the residual r_{1i}, and with $\hat{y}_1$ instead of y_1 to indicate that the response is now a *predicted* quantity:

$$\hat{y}_1 = \beta_0 + \beta_1 x_1 \tag{9.6}$$

For Figure 9.1, the fitted model is

$$\hat{y}_1 = 15.0 + 0.750 x_1 \tag{9.7}$$

The straight line has an intercept (β_0) of 15.0 (that is, when $x_1 = 0$, $\hat{y}_1 = 15.0$) and a slope with respect to x_1 (β_1) of 0.750. Straight-line models of this form are very widespread.

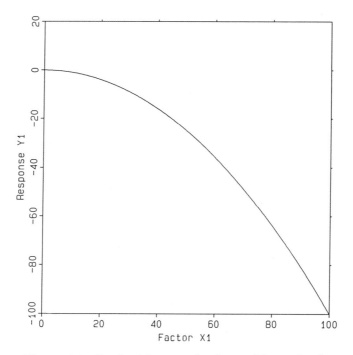

Figure 9.3 Graph of the second-order model $\hat{y}_1 = \beta_{11}x_1^2$.

Figure 9.2 illustrates another straight-line relationship that is slightly differ-ent from that shown in Figure 9.1. The intercept has been forced to zero ($\beta_0 = 0$). This is equivalent to dropping β_0 from the model. The resulting straight line goes through the origin. This probabilistic model is called the "straight line constrained to go through the origin" and is of the form

$$y_{1i} = \beta_1 x_{1i} + r_{1i} \qquad\qquad (9.8)$$

The fitted model that describes the straight-line relationship shown in Figure 9.2 is

$$\hat{y}_1 = 0.750 x_1 \qquad\qquad (9.9)$$

Models of this form are often used by kineticists who believe that no reactant will give no reaction. For example,

$$r = k[X] \qquad\qquad (9.10)$$

where r is the reaction rate (y_1), k is the rate constant (β_1), and $[X]$ is the concentra-tion of reactant X (x_1). Analytical chemists also use this form of the straight-line

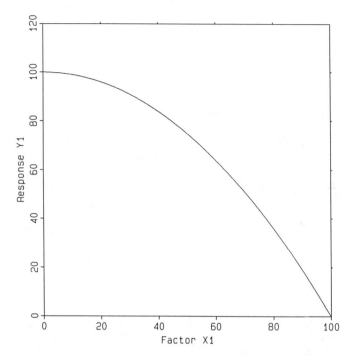

Figure 9.4 Graph of the second-order model with intercept, $\hat{y}_1 = \beta_0 + \beta_{11}x_1^2$.

model for spectrophotometric (and other) calibration curves if a so-called "blank" has been subtracted from all of the other data. For example, Beer's law states that

$$A = \varepsilon bC \tag{9.11}$$

where A is the absorbance (y_1), εb is the product of molar absorptivity ε times the pathlength b (combined to give the slope β_1), and C is the concentration of analyte (substance being determined).

The presence of a β_0 term in a model provides a degree of freedom for "correcting" or adjusting the data for the mean in the analysis of variance; the absence of a β_0 term thus prevents the "correction" or adjustment of data for the mean. Our preference is to initially include the β_0 term in the model—after all, sometimes reactions do occur without the supposed reactant, and sometimes blanks are measured incorrectly. If β_0 is estimated sufficiently close to zero then it might be dropped from the model and the model refit [1].

Straight-line models are often sufficient approximations to the true behavior of the system far from the optimum (e.g., on the sides of hills), but straight-line models are not very good for describing response surfaces in the region of the optimum. This is because the region of the optimum usually shows curvature that cannot be explained by straight-line relationships. Curvature is often accounted for by higher order terms (usually squared terms) in linear models. A simple form of a second-order model is

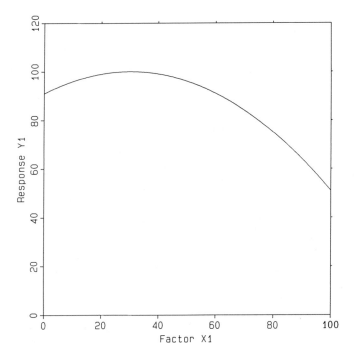

Figure 9.5 Graph of the full one-factor second-order model $\hat{y}_1 = \beta_0 + \beta_1 x_1 + \beta_{11} x_1^2$.

$$y_{1i} = \beta_{11} x_{1i}^2 + r_{1i} \tag{9.12}$$

The parameter β_{11} is usually negative to give a downward-sloping parabolic shape, thus allowing the function to go through a maximum. (Clearly, if the optimum were represented by a minimum, then β_{11} would be positive to give a parabola that opens upward.)

Figure 9.3 is a graph of a very simple second-order model:

$$\hat{y}_1 = -0.0100 x_1^2 \tag{9.13}$$

Note that the vertical axis in Figure 9.3 goes from 20 at the top to -100 at the bottom. Because most maximized responses are not negative, Equation 9.13 is not a very useful form of the model. Instead, the model can be "lifted" (or, equivalently, the y_1 axis can be "lowered") by adding a β_0 term to the model:

$$y_{1i} = \beta_0 + \beta_{11} x_{1i}^2 + r_{1i} \tag{9.14}$$

Figure 9.4 shows the fitted model

$$\hat{y}_1 = 100.0 - 0.0100 x_1^2 \tag{9.15}$$

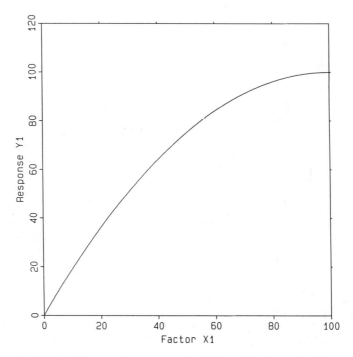

Figure 9.6 Graph of the one-factor second-order model constrained to go through the origin, $\hat{y}_1 = \beta_1 x_1 + \beta_{11} x_1^2$.

Note that the vertical axis in Figure 9.4 goes from 120 at the top to zero at the bottom.

The "slope" term ($\beta_1 x_1$) in higher order models serves a slightly different function than it does in straight-line models: combined with β_0, its function is to move the point of the optimum away from the origin. Figure 9.5 is a graph of the full one-factor second-order model of general form

$$y_{1i} = \beta_0 + \beta_1 x_{1i} + \beta_{11} x_{1i}^2 + r_{1i} \qquad (9.16)$$

Specifically,

$$\hat{y}_1 = 91.0 + 0.600 x_1 - 0.0100 x_1^2 \qquad (9.17)$$

which can be rewritten

$$\hat{y}_1 = 100.0 - 0.0100 (x_1 - 30.0)^2 \qquad (9.18)$$

to show that when $x_1 = 30.0$, the response $\hat{y}_1$ is predicted to be at its maximum value of 100.0.

Finally, the parabolic model can be forced through the origin by setting $\beta_0 = 0$ (i.e., removing it from the model). Figure 9.6 graphs a model of the form

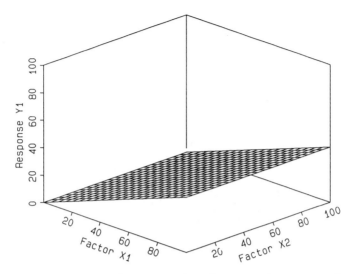

Figure 9.7 The model $\hat{y}_1 = \beta_1 x_1$ plotted over two factors, x_1 and x_2.

$$y_{1i} = \beta_1 x_{1i} + \beta_{11} x_{1i}^2 + r_{1i} \tag{9.19}$$

Specifically,

$$\hat{y}_1 = 2.00x_1 - 0.0100x_1^2 \tag{9.20}$$

Multifactor Systems

The examples in this section will be limited to just two factors, x_1 and x_2. The results are general, however, and may be extended to higher order factor spaces.

Occasionally, one or more of the factors in a multidimensional model will not have a significant effect on the response (i.e., the βs involving that factor will equal zero). When this is the case, that factor can be removed from the model and the dimensionality of the presentation can be decreased by one.

For example, suppose factor x_2 does not exert a significant effect in a two-factor system. If the response *is* a function of the remaining factor x_1, then a model can still be plotted over a two-dimensional factor space. Figure 9.7 is a pseudo-three-dimensional graph of the model

$$\hat{y}_1 = 0.400x_1 \tag{9.21}$$

Note that the model goes through the origin. In fact, the model passes through the intersection of the y_1–x_2 and x_1–x_2 planes. That is, when $x_1 = 0$, $\hat{y}_1 = 0$ for all values of x_2. The model has a slope of 0.400 for x_1 (at all values of x_2). The model has a slope of 0 for x_2 (at all values of x_1).

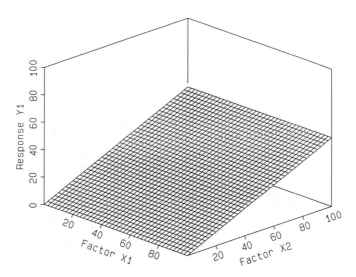

Figure 9.8 The model $\hat{y}_1 = \beta_2 x_2$ plotted over two factors, x_1 and x_2.

Zero slopes like that exhibited by factor x_2 in Figure 9.7 suggest that the associated factor will not have very much influence on the response over the region described by the model. This in turn means that the factor cannot be used to adjust the response to a desired level. However, it also means that unintentional changes in this factor will contribute very little noise to the system. In general, zero slopes are undesirable for process factors; zero slopes are desirable for environmental or "noise" factors.

Figure 9.8 shows a similar two-factor response surface for which the first factor x_1 has no effect ($\beta_1 = 0$). The model is of the form

$$y_{1i} = \beta_2 x_{2i} + r_{1i} \tag{9.22}$$

Specifically,

$$\hat{y}_1 = 0.500 x_2 \tag{9.23}$$

The model has a slope of 0.500 for x_2 (at all values of x_1). The model has a slope of 0 for x_1 (at all values of x_2).

When both factors have an effect on the response, then both factors must be included in the model. If the response surface is constrained to go through the origin, then the two-factor first-order model is simply

$$y_{1i} = \beta_1 x_{1i} + \beta_2 x_{2i} + r_{1i} \tag{9.24}$$

Figure 9.9 plots the specific model

$$\hat{y}_1 = 0.400 x_1 + 0.500 x_2 \tag{9.25}$$

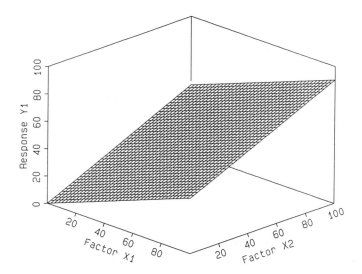

Figure 9.9 Graph of the model $\hat{y}_1 = \beta_1 x_1 + \beta_2 x_2$.

Note that the model has a slope of 0.400 for x_1 (at all values of x_2). The model also has a slope of 0.500 for x_2 (at all values of x_1). The effects are additive. Thus, it is evident from Figure 9.9 that when $x_1 = 100.0$ and $x_2 = 100.0$, $\hat{y}_1 = 0.400 \times 100.0 + 0.500 \times 100.0 = 40.0 + 50.0 = 90.0$.

If the model is not to be constrained to go through the origin, then a β_0 term is added to give the model

$$y_{1i} = \beta_0 + \beta_1 x_{1i} + \beta_2 x_{2i} + r_{1i} \tag{9.26}$$

Figure 9.10 plots the specific model

$$\hat{y}_1 = 8.0 + 0.400 x_1 + 0.500 x_2 \tag{9.27}$$

When interaction is present in multifactor systems, planar models such as Equations 9.24 and 9.26 cannot adequately describe the "twisted planes" that result from the interaction. Instead, a term must be added to the model that will allow the slope with respect to one factor to change as the other factor changes. This is easily done with a $\beta_{12} x_1 x_2$ term to give models of the general form

$$y_{1i} = \beta_0 + \beta_1 x_{1i} + \beta_2 x_{2i} + \beta_{12} x_{1i} x_{2i} + r_{1i} \tag{9.28}$$

Figure 9.11 plots the specific model

$$\hat{y}_1 = 8.0 + 0.400 x_1 + 0.500 x_2 - 0.00800 x_1 x_2 \tag{9.29}$$

Note that the effect (slope) of each factor depends on the value of the other factor. For example, when $x_2 = 0$, the slope with respect to x_1 is +0.400. But when $x_2 = 100$,

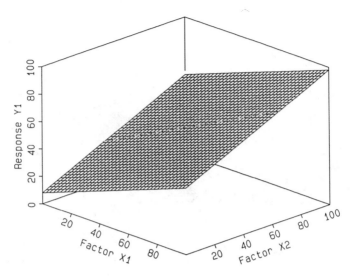

Figure 9.10 Graph of the model $\hat{y}_1 = \beta_0 + \beta_1 x_1 + \beta_2 x_2$.

interactions, and all 1st order terms

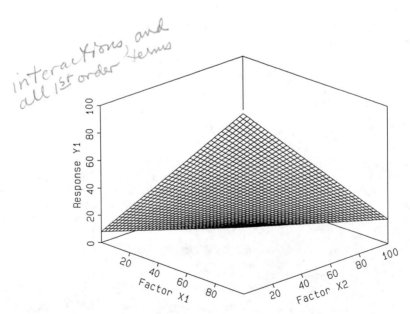

Figure 9.11 Graph of the model $\hat{y}_1 = \beta_0 + \beta_1 x_1 + \beta_2 x_2 + \beta_{12} x_1 x_2$ showing interaction between the factors x_1 and x_2.

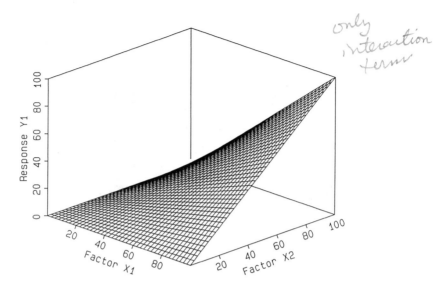

only interaction terms

Figure 9.12 Graph of the model $\hat{y}_1 = \beta_{12}x_1x_2$ showing interaction between the factors x_1 and x_2.

the slope with respect to x_1 is -0.400. Similarly, when $x_1 = 0$, the slope with respect to x_2 is $+0.500$, but when $x_1 = 100$, the slope with respect to x_2 is -0.300. Again, interaction between factors means that the effect of one factor depends on the value of the other factor. In Equations 9.28 and 9.29 it is the $\beta_{12}x_1x_2$ term that accounts for the interaction visible in Figure 9.11.

Figure 9.12 is the graph of a two-factor model that contains an interaction term only:

$$y_{1i} = \beta_{12}x_{1i}x_{2i} + r_{1i} \tag{9.30}$$

Specifically,

$$\hat{y}_1 = 0.0100x_1x_2 \tag{9.31}$$

This type of model is often encountered in kinetic studies involving rate laws of the type

$$r = k[X][Y] \tag{9.32}$$

where, again, r is the reaction rate (y_1), k is the rate constant (β_1), and $[X]$ and $[Y]$ are the concentrations of reactants X (x_1) and Y (x_2).

Curvature in a single factor is introduced using a higher order (typically squared) term. For curvature in the x_1 factor, models of the form

$$y_{1i} = \beta_0 + \beta_1x_{1i} + \beta_{11}x_{1i}^2 + r_{1i} \tag{9.33}$$

↑ curvature term

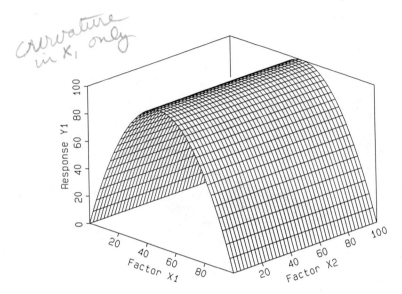

Figure 9.13 The model $\hat{y}_1 = \beta_0 + \beta_1 x_1 + \beta_{11} x_1^2$ plotted over two factors, x_1 and x_2.

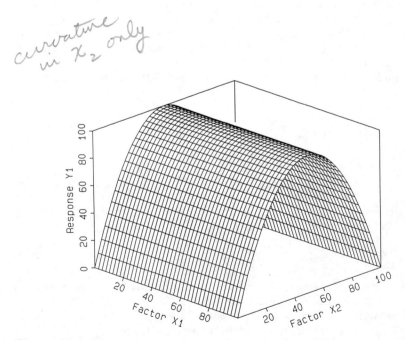

Figure 9.14 The model $\hat{y}_1 = \beta_0 + \beta_2 x_2 + \beta_{22} x_2^2$ plotted over two factors, x_1 and x_2.

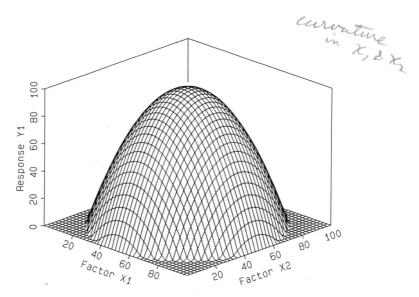

curvature in x_1 & x_2 (handwritten annotation)

Figure 9.15 Graph of the model $\hat{y}_1 = \beta_0 + \beta_1 x_1 + \beta_2 x_2 + \beta_{11} x_1^2 + \beta_{22} x_2^2$.

are useful. Figure 9.13 plots the specific model

$$\hat{y}_1 = 0.000 + 4.00 x_1 - 0.0400 x_1^2 \tag{9.34}$$

Similarly, for curvature in the x_2 factor, Figure 9.14 shows an example of the general model

curvature in x_2 only (handwritten annotation)

$$y_{1i} = \beta_0 + \beta_2 x_{2i} + \beta_{22} x_{2i}^2 + r_{1i} \tag{9.35}$$

Specifically, Figure 9.14 plots the fitted model

$$\hat{y}_1 = 0.000 + 4.00 x_2 - 0.0400 x_2^2 \tag{9.36}$$

Curvature in both factors, x_1 and x_2, can be accounted for by combining Equations 9.33 and 9.35 into a model of the form

curvature in both x_1 & x_2 (handwritten annotation)

$$y_{1i} = \beta_0 + \beta_1 x_{1i} + \beta_2 x_{2i} + \beta_{11} x_{1i}^2 + \beta_{22} x_{2i}^2 + r_{1i} \tag{9.37}$$

Figure 9.15 plots the specific model

$$\hat{y}_1 = -100.0 + 4.00 x_1 + 4.00 x_2 - 0.0400 x_1^2 - 0.0400 x_2^2 \tag{9.38}$$

where the curvatures in x_1 and x_2 are the same. When the extent of curvature is different, the figure becomes elongated in one factor dimension relative to the other factor dimension. Figure 9.16 plots the fitted model

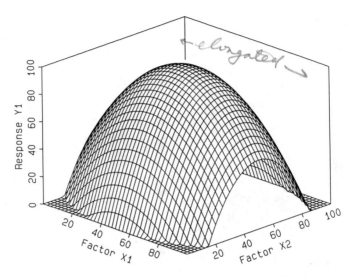

Figure 9.16   Graph of the model $\hat{y}_1 = \beta_0 + \beta_1 x_1 + \beta_2 x_2 + \beta_{11} x_1^2 + \beta_{22} x_2^2$, where $|\beta_{11}| < |\beta_{22}|$.

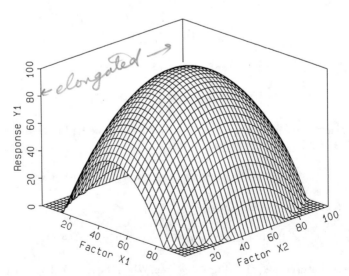

Figure 9.17   Graph of the model $\hat{y}_1 = \beta_0 + \beta_1 x_1 + \beta_2 x_2 + \beta_{11} x_1^2 + \beta_{22} x_2^2$, where $|\beta_{11}| > |\beta_{22}|$.

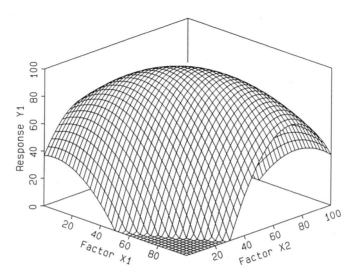

Figure 9.18 Graph of the full two-factor second-order polynomial model $\hat{y}_1 = \beta_0 + \beta_1 x_1 + \beta_2 x_2 + \beta_{11} x_1^2 + \beta_{22} x_2^2 + \beta_{12} x_1 x_2$.

$$\hat{y}_1 = -50.0 + 2.00x_1 + 4.00x_2 - 0.0200x_1^2 - 0.0400x_2^2 \tag{9.39}$$

for which the curvature is greater in x_2 than it is in x_1. Figure 9.17 plots the fitted model

$$\hat{y}_1 = -50.0 + 4.00x_1 + 2.00x_2 - 0.0400x_1^2 - 0.0200x_2^2 \tag{9.40}$$

for which the curvature is greater in x_1 than it is in x_2.

When interaction is present in second-order models, it has the effect of rotating the response surface. Figure 9.18 is an example of the general full two-factor second-order polynomial model

$$y_{1i} = \beta_0 + \beta_1 x_{1i} + \beta_2 x_{2i} + \beta_{11} x_{1i}^2 + \beta_{22} x_{2i}^2 + \beta_{12} x_{1i}\, x_{2i} + r_{1i} \tag{9.41}$$

The model is a "full two-factor second-order polynomial model" because it contains all possible polynomial terms in two factors of order two or lower: β_0, β_1, β_2, β_{11}, β_{22}, and β_{12}. The parameter β_0 is a zero-order parameter. The two parameters β_1 and β_2 are first-order parameters. The three parameters β_{11}, β_{22}, and β_{12} are second-order parameters. Figure 9.18 is the graph of

$$\hat{y}_1 = 36.60 + 0.2680x_1 + 2.268x_2 - 0.0200x_1^2 - 0.0400x_2^2 + 0.03464x_1 x_2 \tag{9.42}$$

It must be remembered that models are approximations to the true response surface. As such, they are usually adequate if they are used over a limited factor domain. As the size of the domain increases, the greater is the chance that the

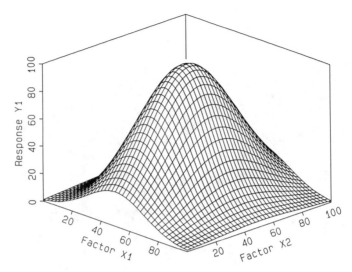

Figure 9.19 Graph of the nonlinear model $\hat{y}_1 = \beta_0 \exp[-(x_1-\beta_1)^2/(2.00\beta_2^2)]$ $\exp[-(x_2-\beta_3)^2/(2.00\beta_4^2)]$.

model will begin to exhibit some lack of fit. Figure 9.19 represents a response surface that is probably typical of most real response surfaces. Over limited subregions of Figure 9.19, the models that have been presented up to this point will be adequate. To model the whole surface, however, more complex models must be used. These more comprehensive models are usually mathematically nonlinear (Chapter 10). For example, the response surface shown in Figure 9.19 is a two-factor multiplicative gaussian of the form

$$y_{1i} = \beta_0 \exp[-(x_{1i}-\beta_1)^2/(2.00\beta_2^2)]\exp[-(x_{2i}-\beta_3)^2/(2.00\beta_4^2)] + r_{1i} \qquad (9.43)$$

Specifically,

$$\hat{y}_1 = 100.0 \times \exp[-(x_1-50)^2/(800)] \times \exp[-(x_2-50)^2/(1800)] \qquad (9.44)$$

Taylor's theorem states that any continuous function can be approximated with any degree of precision by a polynomial of sufficiently high order. Thus, the response surface drawn in Figure 9.19 and represented exactly by Equation 9.44 can be approximated with a polynomial model of the form

$$y_{1i} = \beta_0$$
$$+ \beta_1 x_{1i} + \beta_2 x_{2i}$$
$$+ \beta_{11} x_{1i}^2 + \beta_{22} x_{2i}^2 + \beta_{12} x_{1i} x_{2i}$$
$$+ \beta_{111} x_{1i}^3 + \beta_{222} x_{2i}^3 + \beta_{112} x_{1i}^2 x_{2i} + \beta_{122} x_{1i} x_{2i}^2$$
$$+ \beta_{1111} x_{1i}^4 + \cdots + r_{1i} \qquad (9.45)$$

However, the optimum lies in only a small subregion of the entire factor space shown in Figure 9.19. It is not necessary to use a model that will fit the entire factor space, only the small region around the optimum. Terms greater than second-order in the Taylor series approximation to the response surface (Equation 45) can be assumed to be of lesser importance in the region of the optimum, and can be dropped from the model. Thus, in general, a full second-order polynomial model (Equation 9.41) is usually satisfactory for approximating the response surface in the region of the optimum.

The total number of parameters (p) in a full second-order polynomial model in k factors is given by the equation

$$p = \tfrac{1}{2}(k + 1)(k + 2) \tag{9.46}$$

(handwritten annotation: table with columns "Factors | params": 1 | 3; 2 | 6; 3 | 10; 4 | 15)

REGRESSION ANALYSIS

Matrix least-squares is usually used to fit a linear model (such as Equation 9.41) to a set of data. The parameters of the model (the βs) are estimated so that the sum of squares of residuals (SS_r, the sum of the r_{1i}^2s) is minimized. The parameter *estimates* are usually given the symbol b to distinguish them from the parameters themselves. With linear models, the results obtained are the best possible in the least-squares sense: there will be no other combination of parameter estimates that gives a smaller sum of squares of residuals for the chosen model and the given data set.

Matrix least squares is sometimes called "linear least squares." The method cannot be used directly for nonlinear models (see Chapter 10). It is also called "regression analysis" after one of its first applications, a study by Galton of the reversion of population characteristics back to their mean in later generations (see Stigler [2] for a discussion).

Details of regression analysis are beyond the scope of this presentation but may be found in standard texts [1, 3–5]. Many computer programs are available for carrying out the actual calculations.

ANALYSIS OF VARIANCE
FOR LINEAR MODELS

Regression analysis can provide a wealth of other information in addition to the desired parameter estimates. In particular, analysis of variance of linear models provides a partitioning of the total variance into individual contributions that may be combined in various ways to give diagnostic information about how well the model fits the data. Many details of the analysis of variance are beyond the scope of this presentation, but an overview of the additivity of sums of squares and degrees of freedom is helpful to fully understand classical experimental design.

Figure 9.20 is a diagram showing the additivity of sums of squares and degrees of freedom for linear models containing a β_0 term.

The "total sum of squares", $SS_t = \Sigma\, y_{1i}^2$ (where the summation is always over all n data points), can be partitioned into a "sum of squares due to the mean,"

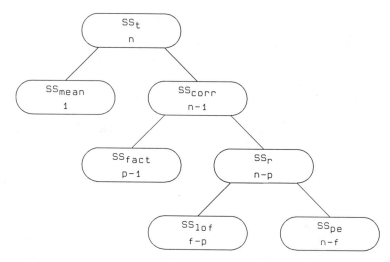

Figure 9.20 The sums of squares and degrees of freedom tree for mathematically linear models containing a β_0 term.

$SS_{mean} = \Sigma \bar{y}_1^2$ (where $\bar{y}_1$ is simply the average of all of the n responses), and a "sum of squares corrected for the mean," $SS_{corr} = \Sigma(y_{1i} - \bar{y}_1)^2$. (The data are adjusted, really, not corrected.)

The sum of squares corrected for the mean can be further partitioned into a "sum of squares due to the factors as they appear in the model" (usually called the "sum of squares due to regression"), $SS_{fact} = SS_{reg} = \Sigma(\hat{y}_{1i} - \bar{y}_1)^2$, and a "sum of squares of residuals," $SS_r = \Sigma(\hat{y}_{1i} - y_{1i})^2$.

Finally, the sum of squares of residuals can be partitioned into a "sum of squares due to lack of fit," $SS_{lof} = \Sigma(\tilde{y}_{1i} - \hat{y}_{1i})^2$, and a "sum of squares due to purely experimental uncertainty" (usually called the "sum of squares due to pure error"), $SS_{pe} = \Sigma(y_{1i} - \tilde{y}_{1i})^2$ (where $\tilde{y}_{1i}$ is the response averaged over a given subset of replicates to which the particular y_{1i} belongs).

For linear models, the sums of squares are additive as shown in Figure 9.20: $SS_{pe} + SS_{lof} = SS_r$, $SS_r + SS_{fact} = SS_{corr}$, and $SS_{corr} + SS_{mean} = SS_t$.

Each sum of squares in Figure 9.20 has associated with it a number of degrees of freedom.

SS_t has n degrees of freedom because at this point each of the n data points can have any value. No restrictions have been placed on the data initially.

SS_{mean} has one degree of freedom because it is based on only one piece of information, the average of all responses, $\bar{y}_1$.

After the mean has been subtracted from all of the original data points, the resulting differences will have $n - 1$ degrees of freedom. This is because the resulting differences must add up to zero, an equality constraint that removes the one degree of freedom. Thus, SS_{corr}, which is calculated from these differences, has $n - 1$ degrees of freedom.

SS_{fact} has $p - 1$ degrees of freedom because, of the p parameters in the model, there are only $p - 1$ factor effects. The parameter β_0 is not associated with any of the factors. Its presence in the model supplies the one degree of freedom required by $\bar{y}_1$.

SS_r has $n - p$ degrees of freedom. To be able to calculate the residuals, the model must have been fitted so responses can be estimated for each data point. But if the model has been fitted, then each of the parameters will have been estimated, and each estimate requires one degree of freedom. Thus, the set of residuals has the original n degrees of freedom minus the p degrees of freedom that have been used to fit the model.

SS_{lof} has $f - p$ degrees of freedom (where f is the number of "distinctly different factor combinations," or "design points," or "treatment combinations," or "treatments," in a set of experiments). The expression, "Any fool can put a straight line through two points" is a reflection of this number of degrees of freedom. If there are two points (presumably two different levels of the factor), then $f = 2$. If there is a straight-line model, then $p = 2$ (β_0 and β_1). Thus, there will be no degrees of freedom for lack of fit ($f - p = 2 - 2 = 0$) and the straight line will fit perfectly. It is only by carrying out experiments at more than p design points that the model is given a chance to disprove itself; otherwise, the model will appear to "fit perfectly."

Finally, SS_{pe} has $n - f$ degrees of freedom. If experiments have been carried out at only f design points, and if there are more than f experiments in the set of n data points, then some of the experiments must have been carried out at design points where other experiments had already been done. In other words, some of the design points must be associated with replicated experiments. It is these "extra" experiments that contribute to an understanding of purely experimental uncertainty, or pure error. If n experiments have been carried out at only f design points, then $n - f$ of them must represent these "extra" experiments.

For linear models, the degrees of freedom are additive as shown in Figure 9.20: $(n - f) + (f - p) = (n - p); (n - p) + (p - 1) = (n - 1); (n - 1) + 1 = n$.

The sums of squares and degrees of freedom tree shown in Figure 9.20 is useful for understanding common regression analysis diagnostics. For any linear model, the "coefficient of multiple determination," R^2, is simply the sum of squares due to the factors as they appear in the model divided by the sum of squares corrected for the mean:

$$R^2 = SS_{fact}/SS_{corr} \qquad (9.47)$$

The square root of this quantity is the "coefficient of multiple correlation," R. When the linear model is the common two-parameter straight-line relationship (Equation 9.3), the same ratio of sums of squares is called the "coefficient of determination," r^2:

$$r^2 = SS_{fact}/SS_{corr} \qquad (9.48)$$

The square root of this quantity is the common "coefficient of correlation," or "correlation coefficient," r.

Statistically, an estimated variance is given the symbol s^2 and is calculated by dividing a sum of squares by its associated degrees of freedom. Thus,

$$s_t^2 = SS_t/n \tag{9.49}$$

$$s_{mean}^2 = SS_{mean}/1 \tag{9.50}$$

$$s_{corr}^2 = SS_{corr}/(n-1) \tag{9.51}$$

$$s_{fact}^2 = SS_{fact}/(p-1) \tag{9.52}$$

$$s_r^2 = SS_r/(n-p) \tag{9.53}$$

$$s_{lof}^2 = SS_{lof}/(f-p) \tag{9.54}$$

$$s_{pe}^2 = SS_{pe}/(n-f) \tag{9.55}$$

Some of these variances can be used to form Fisher variance ratios. One of these calculated ratios is used to test the significance of regression:

$$F_{reg} = F_{(p-1,n-p)} = s_{fact}^2/s_r^2 \tag{9.56}$$

The larger this ratio, the more confident the investigator can be that at least one of the factor effects in the model is probably real and is not caused by chance arrangements in the data set. The ratio calculated using Equation 9.56 is compared with tabulated values to determine the exact level of confidence.

Another Fisher variance ratio is used to test the lack of fit of the model to the data:

$$F_{lof} = F_{(f-p,n-f)} = s_{lof}^2/s_{pe}^2 \tag{9.57}$$

The larger this ratio, the more confident the investigator can be that the model exhibits a lack of fit. By itself, a highly statistically significant lack of fit is not necessarily bad. Even though a model might show a large lack of fit, the model might still be good enough for practical purposes. A highly significant lack of fit simply means that if the investigator wants to look for a better model, it will be possible to recognize such a model once it has been found; if the lack of fit is not very significant, then it probably is not worthwhile looking for a better model—the investigator wouldn't be able to recognize it even if it were found.

All of this prior discussion seems to involve data treatment, not experimental design. But the above discussion is also relevant to experimental design for the following reasons:

1. If the number of degrees of freedom associated with SS_{lof} is negative, the chosen model cannot be fitted to the data set. This means that the number of design points must be greater than or equal to the number of parameters in the model: $f \geq p$. Thus, if the investigator has chosen a model with p parameters, then the experimental design must have at least p design points. If $f = p$, then the model will (at best) appear to give an exact fit; the residuals will (at best) each be equal to zero.
2. If the number of degrees of freedom associated with any sum of squares is zero, then the sum of squares will be equal to zero. Because of this, if the investigator wants to calculate meaningful values for F_{reg}, F_{lof}, and/or R^2, then there must

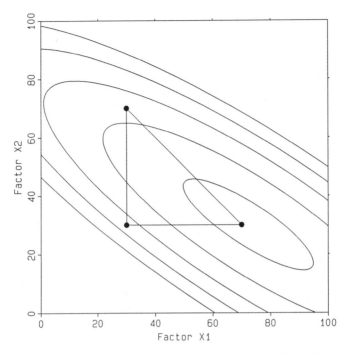

Figure 9.21 An experimental design (simplex) containing three design points.

be degrees of freedom available for each of the sums of squares involved. As suggested by Figure 9.20, this means that n must be greater than f (for SS_{pe}), and f must be greater than p (for SS_{lof}). If these conditions are satisfied, then n will be greater than p (for SS_r).

As a general rule, n should exceed f by 3 to provide 3 or more degrees of freedom for estimating purely experimental uncertainty. Similarly, f should exceed p by 3 to provide 3 or more degrees of freedom for estimating lack of fit. (Some statisticians say 4, others 5, etc., but 3 is probably a minimum difference.) In a very practical sense, this means that

1. the number of design points should be at least 3 more than the number of parameters in the most complicated model that might eventually be fitted to the data; and
2. the total number of experiments should be at least 3 more than the number of design points.

Taken together, these two recommendations suggest that the total number of experiments in an adequate experimental design should be at least 6 more than the number of parameters in the most complicated model that might eventually be fitted to the data.

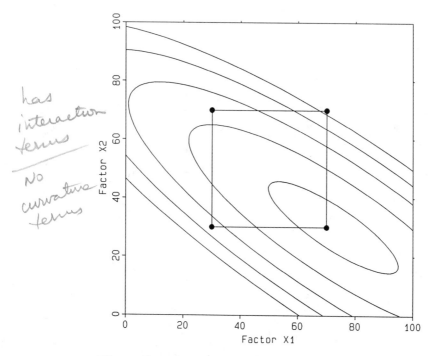

has
interaction
terms
_____
No
curvature
terms

Figure 9.22   A full two-factor two-level (2^2) factorial design.

EXPERIMENTAL DESIGN

Figure 9.21 shows an experimental design that might be used to model a response surface. Because the experimental design has only three design points [(30,30), (30,70), and (70,30)], the most complicated model that could be fitted to the data would have three parameters; otherwise, $f - p$, the number of degrees of freedom for lack of fit, would be negative. Thus, the model given by Equation 9.26 ($y_{1i} = \beta_0 + \beta_1 x_{1i} + \beta_2 x_{2i} + r_{1i}$) might be fitted to the data in Figure 9.21. The model describes a flat plane (Figure 9.10) and would not be able to account for any curvature that might be found in the response surface.

The experimental design in Figure 9.21 is a simplex design. The simplex design is the minimal design required to provide first-order estimates of the factor effects, β_1 and β_2. This type of information can be used for "screening" (sieving) factors to determine which ones probably have the greatest (first-order) effect on the response. In general, large bs will produce large changes in response while small bs will produce small changes in response. Sieving the bs and retaining only those factors that have large estimated factor effects will produce a subset of factors that will probably have the greatest effects in further experimentation. Other screening designs that provide this type of information are saturated fractional factorial designs [6], Plackett-Burman designs [7], and Taguchi designs [8–10].

Figure 9.22 is a two-level two-factor full factorial design. A two-level full factorial design in k factors contains 2^k design points. For two factors, a full factorial

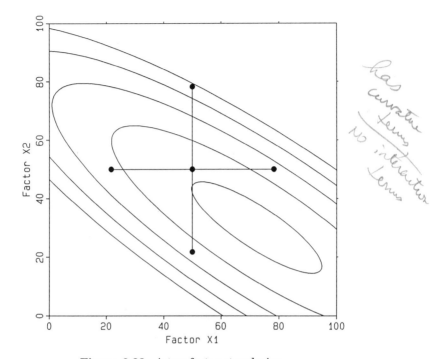

Figure 9.23   A two-factor star design.

design contains $2^2 = 4$ design points. In two factors, data from the full factorial design can be used to estimate the parameters of the model given by Equation 9.28 ($y_{1i} = \beta_0 + \beta_1 x_{1i} + \beta_2 x_{2i} + \beta_{12} x_{1i} x_{2i} + r_{1i}$). Because this model includes an interaction term ($\beta_{12} x_1 x_2$), it can account for twist or rotation in the response surface (see, for example, Figure 9.11) and often does a better job of describing the response surface than does the simpler Equation 9.26.

The experimental design shown in Figure 9.22 does not allow for curvature in each factor by itself. In general, if a model is to account for a qth-order effect in a factor, then the experimental design must have $q + 1$ different levels in that factor. In Figure 9.22 there are only two levels in each factor, so only first-order single-factor effects can be calculated.

Figure 9.23 is a "star design." It is constructed by stepping out a positive amount and a negative amount in each factor from a central point while holding all other factors constant. The star design contains $2k + 1$ design points: 2 points that form the "arms" of the star in each of the k dimensions, plus the one central point. Because there are three different levels in each factor (low, middle, and high; $q + 1 = 3$), this model can be used to estimate second-order single-factor effects (curvature in each factor). Equation 9.37 ($y_{1i} = \beta_0 + \beta_1 x_{1i} + \beta_2 x_{2i} + \beta_{11} x_{1i}^2 + \beta_{22} x_{2i}^2 + r_{1i}$) can be fitted to the data from the experimental design shown in Figure 9.23. Note that the model contains five parameters ($p = 5$). The two-factor star design contains five design points ($f = 5$). Thus, fitting the second-order model represented by Equation 9.37 to data from the star design

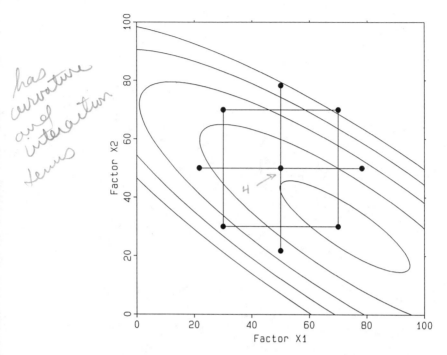

Figure 9.24 A two-factor central composite design.

shown in Figure 9.23 will give an "exact" fit: because $f - p = 5 - 5 = 0$, there will be no degrees of freedom for lack of fit.

Although the star design shown in Figure 9.23 can be used to obtain single-factor curvature effects, it cannot be used to estimate interaction between x_1 and x_2. To estimate interaction, the effect of one factor (e.g., x_1) must be determined at two or more levels of the other factor (x_2). The star design does not allow this.

It is interesting to compare the full factorial design and the star design to see how close each can come to fitting a full two-factor second-order polynomial model. The parameters that can be estimated using each of the designs are as follows:

<table>
<tr><td>2² factorial design:</td><td>$\beta_0 \ \beta_1 \ \beta_2$ — — β_{12}</td></tr>
<tr><td>2-factor star design:</td><td>$\beta_0 \ \beta_1 \ \beta_2 \ \beta_{11} \ \beta_{22}$ —</td></tr>
<tr><td>both designs combined:</td><td>$\beta_0 \ \beta_1 \ \beta_2 \ \beta_{11} \ \beta_{22} \ \beta_{12}$</td></tr>
</table>

In 1951, Box and Wilson [11] proposed the combination of the two designs to generate a composite design that would allow the fitting of a full second-order polynomial model. Because the centers of the two designs coincide, the experimental design is called a "central composite design." A central composite design in two factors is shown in Figure 9.24. The superposition of the 2² factorial design and the 2-factor star design is evident. The number of design points in a k-factor central composite design is $2^k + 2k + 1$ (2^k from the full factorial part of the design, and $2k + 1$ from the star part of the design).

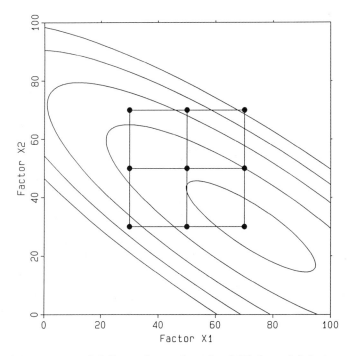

Figure 9.25 A full two-factor three-level (3^2) factorial design.

The center point is usually replicated a total of four times. Thus, the design has nine design points ($f = 9$). The three additional replicates at the center point bring the total number of experiments to 12 ($n = 12$). The number of parameters in the full two-factor second-order polynomial model is six ($p = 6$). From an analysis of variance point of view, this central composite design is a paradigm of what all experimental designs should be:

1. There are three degrees of freedom for assessing purely experimental uncertainty ($n - f = 12 - 9 = 3$).
2. There are three degrees of freedom for assessing lack of fit ($f - p = 9 - 6 = 3$).
3. There are six degrees of freedom for residuals ($n - p = 12 - 6 = 6$).

Thus, R^2 will not equal 1.00 artificially (as it would if there were no degrees of freedom for residuals), F_{reg} will be defined, and F_{lof} can be calculated. Central composite designs are especially appropriate for understanding the response in the region of a suspected optimum. Central composite designs provide sufficient information to fit the full second-order polynomial model with a minimum number of experiments.

Figure 9.25 shows a three-level two-factor full factorial design, a 3^2 design. This can be viewed as a central composite design, where the four corners constitute the 2^2 full factorial and the remaining five points comprise the star design. If the center point is replicated a total of four times, then the analysis of degrees of freedom is identical to that for the central composite design shown in Figure 9.24. In

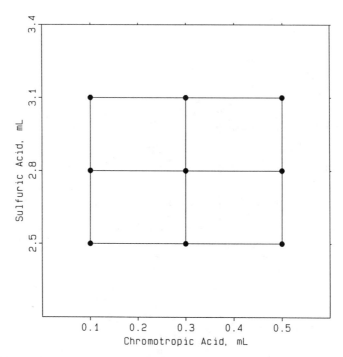

Figure 9.26 The full two-factor three-level (3^2) factorial design used in a study by Olansky and Deming [12].

higher dimensional factor space, the 3^k full factorial design is not equivalent to the k-dimensional central composite design. The 3^k design contains more design points than the central composite design.

Extension of these designs to higher dimensional factor space is straightforward [1, 6].

AN EXAMPLE

The colorimetric determination of formaldehyde was investigated by Olansky and Deming [12]. Two factors were varied: the milliliters (mL) of chromotropic acid (x_1) and the mL of sulfuric acid (x_2). The response (y_1) was the absorbance of the resulting solution after 2.00 mL of a fixed concentration of formaldehyde and the specified volumes of the two reagents were mixed. Simplex optimization was used to increase the absorbance response from 0.080 to 0.599 in 13 experiments (16 vertexes, three of which were boundary violations).

Figure 9.26 shows the full three-level two-factor factorial design ($f = 3^2$ design points) with each point replicated twice ($n = 18$ experiments) that was used to fit a full two-factor second-order polynomial model ($p = 6$ parameters) in the region of the suspected optimum. This was done to verify that an optimum had been reached and to understand the effects of the factors on the response

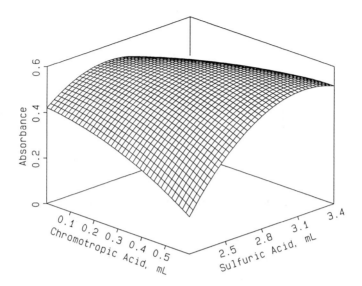

Figure 9.27 Graph of a full two-factor second-order polynomial model used to estimate response in the study by Olansky and Deming [12].

in the region of the optimum. Table 9.1 contains the experimental data, fitted model, and regression diagnostics.

The fitted equation is

$$\hat{y}_1 = -2.615 - 1.479x_1 + 2.321x_2 - 0.525x_1^2 - 0.428x_2^2 + 0.619x_1x_2 \quad \text{(9.58)}$$

R^2 suggests that 80.42% of the variation in the data about its mean is accounted for by the factor effects in the model. F_{reg} (F_{fact}) is significant at the 99.94% level of confidence, which leads to the conclusion that at least one of the factor effects in the model (β_1, β_2, β_{11}, β_{22}, β_{12}) is different from zero. F_{lof} is significant at the 98.81% level of confidence: if we wanted to look for a better model, we would be able to recognize it if we found it.

Figure 9.27 is a pseudo-three-dimensional view of the response surface in the region of the optimum. It shows a diagonal ridge that decreases slightly as both chromotropic acid and sulfuric acid are increased. The chemical reasons why such a ridge might exist are discussed in the original paper [*12*].

Data from the simplex optimization and factorial design studies were combined and used to fit a nonlinear model (using simplex optimization) that was used to provide a more global view of the response surface. The final nonlinear model was

$$\hat{y}_1 = 1.55 \times \exp\{-[(x_2/(x_1 + x_2 + 2.0) - 0.57)^2]/[2.0 \times 0.07^2]\}$$
$$\times [2.0/(x_1 + x_2 + 2.0)] \times [1.0 - \exp(-22 \times x_1)] \quad \text{(9.59)}$$

and is shown in Figure 9.28. It is the subregion at the top of this response surface that is approximated by Equation 9.58 and shown in Figure 9.27. When Equation

Table 9.1 Regression Analysis of Chromotropic Acid Data

Design points 3^2 with replicate

Exp	Factor 1 chromo acid	Factor 2 H_2SO_4	Response	Reps
1	0.1000000	2.5000000	0.5240000	a
2	0.1000000	2.5000000	0.5380000	a
3	0.1000000	2.8000000	0.5150000	b
4	0.1000000	2.8000000	0.5160000	b
5	0.1000000	3.1000000	0.5260000	c
6	0.1000000	3.1000000	0.5300000	c
7	0.3000000	2.5000000	0.4550000	d
8	0.3000000	2.5000000	0.5090000	d
9	0.3000000	2.8000000	0.5830000	e
10	0.3000000	2.8000000	0.5750000	e
11	0.3000000	3.1000000	0.5340000	f
12	0.3000000	3.1000000	0.5450000	f
13	0.5000000	2.5000000	0.3860000	g
14	0.5000000	2.5000000	0.4280000	g
15	0.5000000	2.8000000	0.5450000	h
16	0.5000000	2.8000000	0.5370000	h
17	0.5000000	3.1000000	0.5540000	i
18	0.5000000	3.1000000	0.5510000	i

Parameter estimates   (Determinant = 5.572563D-05)

b		Estimate	% Confidence	Risk
b_0	0	−2.6147222	95.89	0.0411
b_1	1	−1.4791667	99.09	0.0091
b_2	2	2.3210417	98.51	0.0149
b_{11}	11	−0.5250000	86.49	0.1351
b_{22}	22	−0.4277778	98.76	0.0124
b_{12}	12	0.6187500	99.83	0.0017

Same results from Statgraphics

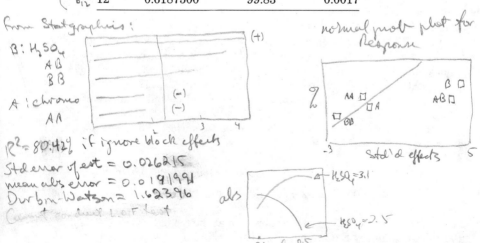

From Statgraphics:

B: H_2SO_4
AB
BB

A: chromo
AA

(+)
(−)
(−)
3   4

normal prob plot for Response

BB   A   AA   BD   ABD

-3   Std'd effects   5

$R^2 = 80.42\%$ if ignore block effects
Std error of est = 0.026215
mean abs error = 0.0191991
Durbin-Watson = 1.62396
Cannot conduct L.O.F. test

als

$H_2SO_4=3.1$
$H_2SO_4=2.5$

.1   chromo   0.5

Breakdown for sums of squares

Exp	Response	Adjusted	Predicted	Residual	Lack of fit	Pure "error"	Reps
1	0.5240000	0.0045000	0.5157917	0.0082083	0.0152084	−0.0070000	a
2	0.5380000	0.0185000	0.5157917	0.0222083	0.0152084	0.0070000	a
3	0.5150000	−0.0045000	0.5505000	−0.0355000	−0.0350000	−0.0005000	b
4	0.5160000	−0.0035000	0.5505000	−0.0345000	−0.0350000	0.0005000	b
5	0.5260000	0.0065000	0.5082083	0.0177917	0.0197917	−0.0020000	c
6	0.5300000	0.0105000	0.5082083	0.0217917	0.0197917	0.0020000	c
7	0.4550000	−0.0645000	0.4873333	−0.0323333	−0.0053333	−0.0270000	d
8	0.5090000	−0.0105000	0.4873333	0.0216667	−0.0053333	0.0270000	d
9	0.5830000	0.0635000	0.5591667	0.0238333	0.0198333	0.0040000	e
10	0.5750000	0.0555000	0.5591667	0.0158333	0.0198333	−0.0040000	e
11	0.5340000	0.0145000	0.5540000	−0.0200000	−0.0145000	−0.0055000	f
12	0.5450000	0.0255000	0.5540000	−0.0090000	−0.0145000	0.0055000	f
13	0.3860000	−0.1335000	0.4168750	−0.0308750	−0.0098750	−0.0210000	g
14	0.4280000	−0.0915000	0.4168750	0.0111250	−0.0098750	0.0210000	g
15	0.5450000	0.0255000	0.5258333	0.0191667	0.0151667	0.0040000	h
16	0.5370000	0.0175000	0.5258333	0.0111667	0.0151667	−0.0040000	h
17	0.5540000	0.0345000	0.5577917	−0.0037917	−0.0052917	0.0015000	i
18	0.5510000	0.0315000	0.5577917	−0.0067917	−0.0052917	−0.0015000	i

Source	Sum of squares	Variance	df
t	4.8999690	0.2722205	18
mean	4.8578445	4.8578445	1
corr	0.0421245	0.0024779	17
fact	0.0338778	0.0067756	5
r	0.0082467	0.0006872	12
lof	0.0056712	0.0018904	3
pe	0.0025755	0.0002862	9

Determination and correlation

R^2 value = 0.8042   *SG gave 80.42% if ignore block effects*

R value = 0.8968

Fisher F-ratios	Estimate	% Confidence	Risk
Fact $F(5,12)$ =	9.859	99.94	0.0006
lof $F(3,9)$ =	6.606	98.81	0.0119

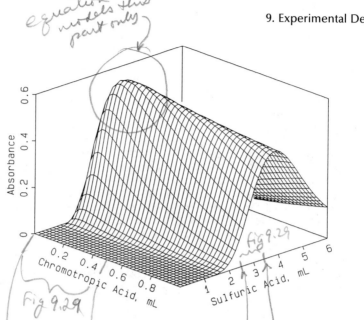

equation 9.5† models this part only

Fig 9.29

Figure 9.28 Graph of the nonlinear model used to estimate response. Adapted from Olansky and Deming [12].

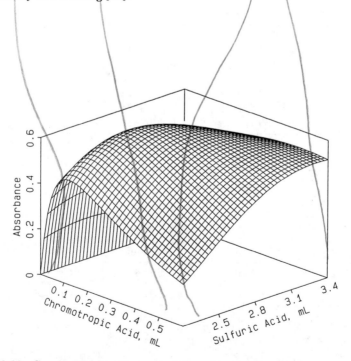

Figure 9.29 Graph of the nonlinear model used to estimate response in the study by Olansky and Deming [12] over the same factor domain as that shown in Figure 9.27.

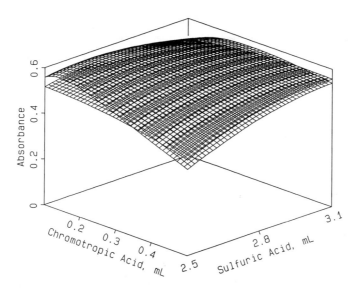

Figure 9.30 Superimposed linear and nonlinear response surfaces over the factor domain covered by the experimental design (Figure 9.26).

9.59 is drawn over the same domain as shown in Figure 9.27, the view shown in Figure 9.29 results (compare with Figure 9.27). Finally, to show the similarity of the two models over the domain covered by the experimental design (Figure 9.26), Figure 9.30 plots the superposition of the two fitted models. Thus, over this limited region of factor space, the linear full two-factor second-order polynomial model does give a good approximation of the more comprehensive nonlinear surface.

Many of the papers on simplex optimization have employed experimental designs in the region of the optimum. These designs have included fractional factorial designs [13], factorial designs [14], replicated factorial designs [15], central composite designs [16], and Box–Behnken designs [17]. The texts by Wheeler are especially well written [18, 19].

REFERENCES

1. S. N. Deming and S. L. Morgan, *Experimental Design: A Chemometric Approach*, Elsevier, Amsterdam, 1987.
2. S. M. Stigler, *The History of Statistics: The Measurement of Uncertainty before 1900*, Harvard University Press, Cambridge, MA, 1986.
3. N. R. Draper and H. Smith, *Applied Regression Analysis*, 2nd ed., Wiley, New York, 1981.
4. O. J. Dunn and V. A. Clark, *Applied Statistics: Analysis of Variance and Regression*, 2nd ed., Wiley, New York, 1987.

5. J. Neter, W. Wasserman, and M. H. Kutner, *Applied Linear Statistical Models: Regression, Analysis of Variance, and Experimental Designs*, 3rd ed., Irwin, Homewood, IL, 1990.
6. G. E. P. Box, W. G. Hunter, and J. S. Hunter, *Statistics for Experimenters: An Introduction to Design, Data Analysis, and Model Building*, Wiley, New York, NY, 1978.
7. R. L. Plackett and J. P. Burman, "The design of optimum multifactorial experiments," *Biometrika*, **33**, 305–325 (1946).
8. A. Bendell, J. Disney, and W. A. Pridmore, Eds., *Taguchi Methods: Applications in World Industry*, IFS Publications, Springer-Verlag, London, 1989.
9. P. J. Ross, *Taguchi Techniques for Quality Engineering: Loss Function, Orthogonal Experiments, Parameter and Tolerance Design*, McGraw-Hill, New York, 1988.
10. G. Taguchi, *Introduction to Quality Engineering: Designing Quality into Products and Processes*, Kraus International Publications, White Plains, NY, 1986.
11. G. E. P. Box and K. B. Wilson, "On the experimental attainment of optimum conditions," *J. R. Stat. Soc.*, **13**, 1–45 (1951).
12. A. S. Olansky and S. N. Deming, "Optimization and interpretation of absorbance response in the determination of formaldehyde with chromotropic acid," *Anal. Chim. Acta*, **83**, 241–249 (1976).
13. L. R. Parker, Jr., S. L. Morgan, and S. N. Deming, "Simplex optimization of experimental factors in atomic absorption spectrometry," *Appl. Spectrosc.*, **29**, 429–433 (1975).
14. S. L. Morgan and S. N. Deming, "Optimization strategies for the development of gas-liquid chromatographic methods," *J. Chromatogr.*, **112**, 267–285 (1975).
15. S. L. Morgan and C. A. Jacques, "Response surface evaluation and optimization in gas chromatography," *J. Chromatogr. Sci.*, **16**, 500–505 (1978).
16. M. L. H. Turoff and S. N. Deming, "Optimization of the extraction of iron(II) from water into cyclohexane with hexafluoroacetylacetone and tri-*n*-butyl phosphate," *Talanta*, **24**, 567–571 (1977).
17. A. S. Olansky, L. R. Parker, Jr., S. L. Morgan, and S. N. Deming, "Automated development of analytical chemical methods. The determination of serum calcium by the cresolphthalein complexone method," *Anal. Chim. Acta*, **95**, 107–133 (1977).
18. Wheeler, Donald J., *Understanding Industrial Experimentation*, Statistical Process Controls, Inc., Knoxville, TN, 1987.
19. Wheeler, Donald J., *Tables of Screening Designs*, 2nd ed., Statistical Process Controls, Inc., Knoxville, TN, 1989.

Chapter **10**

Applications in Chemistry

Because we are analytical chemists, our interests in applications of the sequential simplex have been primarily chemical in nature. This chapter reviews representative applications of the sequential simplex in chemistry and related fields. Unless noted otherwise, applications use the variable-size simplex. A more complete bibliography is given in Chapter 11.

APPLICATIONS IN ANALYTICAL CHEMISTRY

The field of analytical chemistry has many applications of sequential simplex optimization. Analytical chemists are often concerned with developing methods that have maximum sensitivity and minimum interference.

Colorimetric Methods

Colorimetric or spectrometric methods are well suited to optimization: the factors and response are usually straightforward and the sensitivity is easy to determine. Long [1] used simplex to optimize the p-rosaniline determination of sulfur dioxide. A fixed-size simplex was used. After it had located the region of the optimum, a smaller fixed-size simplex was used to further refine the location of the optimum (Chapter 5). Houle et al. [2] applied simplex optimization to the optimization of sensitivity in the colorimetric determination of formaldehyde using chromotropic acid reagent. Because the sample volume was included as a factor and allowed to vary, sensitivity was not actually optimized. Olansky and Deming [3] reinvestigated this method with simplex optimization, keeping the sample volume constant to more accurately optimize sensitivity. In two papers [4, 5],Czech used simplex optimization on the J-acid method and the acetylacetone method for the determination of formaldehyde. Again, sample volume was a factor so sensitivity was not optimized. The sensitivity of the acetylacetone method for formaldehyde was later optimized using simplex by Deming and King [6].

Other colorimetric methods which were optimized include the Liebermann–Burchard method for the determination of cholesterol [7, 8], an extraction method for the determination of phosphate using molybdate [9], the titanium-EDTA method for hydrogen peroxide [10], the synergistic extraction of iron(II) using hexafluoroacetyl acetone and tri-n-butyl phosphate [11], the synergistic extraction of a copper(II) complex [12], a spot test for vanadium [13], and the determination of zinc using 8-hydroxyquinoline [14].

Automated Methods

Automated analyzer methods have been optimized using simplex. One such study involving dialysis [15] optimized the cresolphthalein complexone method for the determination of serum calcium. Six factors (hydrochloric acid concentration before and after dialysis, 8-hydroxyquinoline concentration before and after dialysis, cresolphthalein concentration, and diethylamine concentration) and three responses (the sensitivity to calcium and the effect of two interfering substances, magnesium and protein) were investigated. The objective function was defined as the sensitivity to calcium minus the sensitivity to magnesium minus the protein effect. In addition, boundaries were placed on the responses (Chapter 8): if the magnesium sensitivity or the protein effect was greater than 10 % of the calcium sensitivity, or if the baseline absorbance was greater than the absorbance of the sample highest in calcium, a boundary violation was considered to have occurred and an unfavorable response was assigned to that vertex. Each factor, or reagent, was controlled by a stepper motor with the system under computer control. The computer set the experimental conditions and collected the data. It then subtracted the baseline from all the sample peak absorbances, fit first-order (linear) models to the samples that had varying levels of calcium, magnesium, and protein to obtain the slopes (the sensitivities), computed the objective function from these individual sensitivities (responses), and used the simplex algorithm to determine the location of the next set of experimental conditions. The underlying strategy and philosophy were very similar to what is now known as Taguchi designs [16–18]. An experimental design was carried out in the region of the optimum and a model was fit to better understand the effect of the factors on the primary response (sensitivity) and the two interfering substances. The entire optimization procedure was automated and under computer control. Simplex optimization yielded a combination of factor levels that gave 8.5% greater calcium sensitivity and a baseline 15% lower in absorbance than the standard method.

Other automated clinical chemistry methods investigated using the simplex have included the determination of copper and glucose [19], an enzymatic method for glucose [20], the azomethine-H method for boron [21], and a kinetic method for creatinine [22].

Another closed-loop automated system that was optimized by simplex [23] was a chemical reactor system that used pumps to deliver reagents, valves to empty reservoirs, and temperature and stirrer controls; a sample line led to a liquid chromatograph to measure the yield of product.

A robotic system [24] was used to optimize the indirect spectrophotometric determination of calcium by the indicator Calcon. The robot included a robotic arm, a general-purpose hand, a syringe hand, a liquid sample dispensing station, a vortex mixer, an analog-to-digital converter, and a power and event controller; all of these components were interfaced to a computer. Both fixed- and variable-size simplex algorithms were used.

Optimization of Analytical Instruments

A variety of instruments have been optimized using simplex. Ernst [25] was one of the first to apply simplex optimization to an instrument. He used the algorithm to optimize the homogeneity of the magnetic field of a nuclear magnetic resonance (NMR) spectrometer by varying the current supplied to the linear and quadratic y-gradient magnetic shim coils. This was also one of the first applications using a computer-controlled instrument: a computer (a Digital Equipment Corporation PDP-8) performed the simplex calculations, changed the shim coil currents using stepper motors attached to potentiometers, acquired the response (the peak height of an 8-Hz-wide water line), and then calculated new conditions using the simplex algorithm. The simplex algorithm was found to converge to the region of optimum magnetic field homogeneity in fewer experiments than a modified steepest-descent algorithm. Other applications involving NMR spectrometry optimized the phase and length of pulses in Fourier-transform (FT) NMR [26], the NMR field homogeneity [27], and the phasing of FT NMR spectra [28].

Atomic absorption spectroscopy was the subject of a simplex optimization by Parker et al. [29]. In addition to the expected factors of fuel and oxidant flow rates and burner height, a nonsignificant factor (volume of water in a graduated cylinder) was deliberately included. A one-third fractional three-level factorial design was carried out in the region of the optimum; regression and analysis of variance were performed. Both water volume and lamp current turned out to be nonsignificant over the domains that were varied. Other applications of simplex to atomic absorption spectroscopy have included the optimization of pulsed hollow-cathode lamps [30], the optimization of electrodeless discharge lamps [31, 32], the determination of calcium [33], and general instrument optimization [34–36].

Another atomic spectroscopic technique that has seen several applications of simplex is inductive-coupled plasma (ICP) spectroscopy. Ebdon et al. [37], Ebdon [38], Cave et al. [39], Terblanche et al. [40], Leary et al. [41], Brocas [42], and Gonzalez et al. [43, 44] have optimized various ICP methods using simplex algorithms. Parker et al. [45] used simplex to optimize an ICP method using hydride generation to determine arsenic and selenium. This study optimized in two separate phases the instrumental parameters (outer gas flow rate, intermediate gas flow rate, carrier gas flow rate, and RF power) and the chemical parameters for hydride generation (quantity of acid and quantity of sodium borohydride). The simplex optimization was followed by a mapping study and model fitting. Parker et al. [46], in evaluating simplex algorithms, optimized ICP methods for the determination of copper and manganese. Kornblum et al. [47] compared simplex and univariate methods for optimizing an ICP spectrometer. Cave et al. [48] used the supermodified simplex to determine the optimal operating conditions for 13 elements; a later paper by Barnes and Fodor [49] used these values. Other applications to ICP have also been reported [50, 51].

Other spectrometric methods optimized have included DC plasma arc spectroscopy [52], differential spectroscopy for the determination of chloracizin [53] and for the determination of dimecarbine [54], X-ray fluorescence [55], X-ray radiometric analysis [56], X-ray diffraction of asbestos [57], chemiluminescence [58], fluorescence detection [59], and capacitance-coupled microwave plasma spectrometry [60]. Spaink et al. [61] used simplex to optimize the optical alignment of a tunable diode laser spectrometer; alignment was achieved by displacing a collimation lens placed in front of the laser in three mutually perpendicular directions by stepper motors. The detector signal after lock-in was taken as the response. Simplex has also been used to control a monochromator [62], for flow-injection analysis [63], and to tune up a Fourier transform ion cyclotron resonance mass spectrometer [64].

Chromatography, both gas and liquid, has been optimized using simplex. Gas chromatography was investigated by Morgan and Deming [65]; temperature and carrier gas flow rates were the factors and a chromatography response function (CRF) that took into account the separation of adjacent pairs of peaks was used as the response. The GC system was optimized for two-, three-, and five-component mixtures. A boundary was set on the response: if the total elution time was too long, this was considered a boundary violation and given an undesirable response. A second paper by the same authors [66] discussed optimization in chromatography in a more general sense. Other optimizations involving GC include Holderith et al. [67], Medyantsev et al. [68], Morgan and Jacques [69], Singliar and Koudelka [70], and Sakjartova et al. [71]. Yang et al. [72] optimized capillary GC; Hsu et al. [73] optimized the operation of a GC detector. A column liquid chromatography eluting agent's composition was optimized by Svoboda [74]. Simplex was applied to high-performance liquid chromatography (HPLC) by Rainey and Purdy [75] for the determination of lipids, by Watson and Carr [76] using gradient elution, by Berridge [77–79], and by Berridge and Morrisey [80]. Tomas et al. [81] also applied simplex optimization to chromatographic separations. Walters and Deming [82] compared simplex optimization to window diagrams, another technique for determining optimal mobile phase composition in liquid chromatography. A recent volume by Glajch and Snyder [83] reviews other applications in chromatography.

APPLICATIONS IN OTHER
BRANCHES OF CHEMISTRY

The use of simplex optimization to improve the yield in a synthesis is a natural application, since the factors involved are usually easy to identify and adjust (e.g., pH, temperature, amount of catalyst, amount of reagents) and the response (% yield) is relatively easy to measure. Kofman et al. [84] optimized the synthesis of prometrin. Dean et al. [85] optimized the yield of an organometallic synthesis using the factors reaction time and reaction temperature. Whereas most chemists would run such a reaction by refluxing at the boiling point of the solvent (65 °C for tetrahydrofuran), the yield there with a 1.5-hour reaction time was only 11%; at the optimum (approximately 55°C and 5 hours reaction time), the yield improved to 93%. Dumenil et al. [86] optimized the conversion of DL-homoserine to L-threonine, Dillon and Underwood [87] investigated the formation of allenes, Gilliom et al. [88] optimized drug design, Lazaro et al. [89, 90] optimized the industrial synthesis of pyrazolone, Amenta et al. [91] applied simplex to the synthesis of sec-butyl benzene, Brunel et al. [92] investigated the synthesis of perfluorosulfonic acids, Li-Chan et al. [93] looked at lysine-wheat gluten synthesis, Chubb et al. [94] optimized a Bucherer–Bergs synthesis reaction; and Minkova and Baeva [95] applied simplex to isomerization.

Darvas [96] applied simplex optimization to the design of new drug analogs. Although this work did not attempt to improve the yield, it did use simplex optimization to predict which analogs would have a certain activity. The factors employed here, the Hammett-σ and Hansch-π parameters, were not truly continuous – if the simplex called for a drug molecule with certain values of these factors, substituents were added to the base molecule to come close to the values the simplex gave, but were usually not exactly the same. This led to some distortion of the simplex as it progressed, but its performance was still satisfactory. This type of optimization was continued in a later work by Darvas et al. [97].

Some of the earliest applications of simplex optimization were in industry. Baasel [98] and Carpenter and Sweeny [99] used simplex optimization on industrial processes; Kenworthy [100] applied it in the paper industry. These early optimizations used the fixed-size simplex algorithm. Avots et al. [101] used simplex to optimize a heterogeneous catalyst; Antonenkov and Pomerantsev [102] also applied simplex optimization to a catalyst. Tymczynski et al. [103] used simplex to optimize a rubber-blend composition.

Brusset et al. [104] looked at the oxidation of sulfur dioxide using simplex optimization. Reiss and Katz [105] optimized the recovery of blood platelets from plasma; Silver [106] investigated equilibria involving plutonium. Ceynowa and Wodzki [107] determined the optimal construction of carbon electrodes for use in fuel cells. Evans et al. [108] looked at mannitol-borate interaction. Davydov and Naumov [109], Burgess [110], and Burgess and Hayumku [111] applied simplex to neutron activation analysis; Hanafey et al. [112] applied it to the determination of electrochemical mechanisms. Vlacil and Huynh [113] optimized an analytical procedure for the determination of dibenzyl sulfoxide and Bolanca and Golubovic [114] optimized a procedure for the analysis of ink. Several studies used simplex optimization for kinetics [115–118]. Mayne [119] optimized the preparation of pharmaceutical tablets and Shek et al. [120] optimized the preparation of capsules. In a geochemical application, Kostal et al. [121] used simplex optimization to characterize the composition of geological samples.

NUMERICAL APPLICATIONS

Simplex optimization can be used to fit nonlinear models to data. It was for this purpose that Nelder and Mead [122] modified the fixed-size simplex of Spendley et al. [123]. The "factors" are the coefficients or parameters in the model (the βs) and the response is the sum of squares of residuals, SS_r:

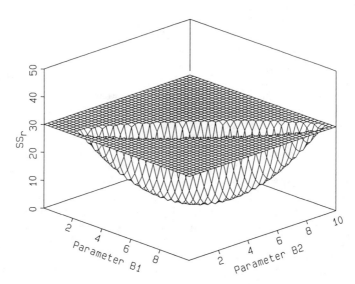

Figure 10.1 The sum of squares of residuals plotted as a function of two parameters for a mathematically linear model. The surface has been truncated by the plane intersecting at $SS_r = 30$; values greater than 30 have been set equal to 30. In general, such surfaces are hyperdimensional paraboloids with elliptical cross sections.

$$SS_r = \sum_{i=1}^{n} (y_{i,\text{observed}} - y_{i,\text{calc}})^2 \qquad (10.1)$$

where $y_{i,\text{calc}}$ is the value predicted by the model at a given factor combination i, $y_{i,\text{observed}}$ is the value actually observed, and n is the number of data points. The object is to find the set of parameter estimates that gives the minimum sum of squares of residuals.

Figures 10.1 and 10.2 illustrate the fundamental difference between linear least squares and nonlinear least squares. With mathematically linear models (models that are first-order in the parameters), the sum of squares surface is a hyperdimensional paraboloid with a single, well-defined minimum (as suggested by Figure 10.1 for two parameters). Linear algebra leads directly to this optimum combination of parameter estimates. There is no guessing, there is only one optimum.

With mathematically nonlinear models (models that are *not* first-order in the parameters), the sum of squares surface is generally bumpy or dimpled, with the possibility of several local minima (as suggested by Figure 10.2 for two parameters). A minimum must be found using iterative trial-and-error search techniques (e.g., the sequential simplex) or approximation techniques (e.g., the paper by Wentworth [124]). The optimum obtained often depends very heavily on the initial guess of the parameter estimates and on the persistence and cleverness of the search algorithm. For a more detailed discussion, see Draper and Smith [125].

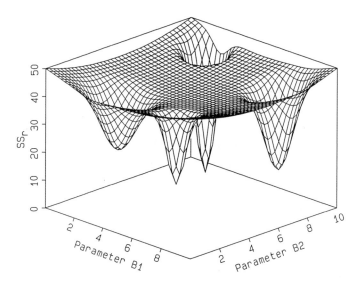

Figure 10.2 The sum of squares of residuals plotted as a function of two parameters for a mathematically nonlinear model (artist's interpretation). Nonlinear models have the potential for possessing multiple local minima in the sums of squares surface.

O'Neill [126] was one of the first to use the simplex algorithm for regression, or model fitting. Caceci et al. [127] demonstrated how a PASCAL computer program could use the simplex algorithm for curve fitting. A PASCAL algorithm is also given by Press et al. [128]. Jurs [129] presented a FORTRAN computer program that used simplex for nonlinear least squares, or regression. The example given used the absorbance data of Deming and Morgan [7] and fit the equation:

$$A = A_\infty (1 - e^{-kt}) \tag{10.2}$$

Delaney and Mauro [130, 131] used simplex optimization to help resolve infrared spectra of mixtures. Normally, surface contouring is used. However, visual inspection of contours in more than two dimensions is difficult, whereas simplex can be used in any number of dimensions. In addition, the use of simplex eliminates uncertainties caused by resolution constraints in gridding and contouring. A three-component mixture of similar compounds (1-hexanol, 1-octanol, and 1-heptanol) was used. The vapor-phase IR spectra from a library were digitized. The simplex correctly, rapidly, and efficiently made identification of the components possible. When one of the components was deleted from the library of spectra, the simplex algorithm had a more difficult time. Dauwe et al. [132] analyzed double decay spectra using the simplex algorithm. Akitt [133] also analyzed spectra, Geiss and Huang [134] analyzed X-ray energy dispersions, Moraweck et al. [135] investigated X-ray line profiles, Fiori and Myklbust [136] used simplex to fit curves to X-ray spectra, Lam et al. [137] used simplex in the deconvolution of spectra, Silber et al. [138] used simplex to fit curves to Mossbauer spectra, Cheng [139] used simplex in NMR calculations, and Leggett [140] applied simplex to multicomponent spectra.

Madsen and Robertson [141] used simplex for nonlinear regression calculations on drug-protein binding parameters. Martin and Hackbarth [142] used simplex for nonlinear regression, as did Van de Voort et al. [143]. Pedersen [144] fit curves to pharmaceutical data using simplex; Kohn et al. [145] fit enzymatic rate equations. Just et al. [146] fit kinetic models for dimerization reactions.

Becker et al. [147] applied simplex to molecular orbital calculations. Berger et al. [148] used it to calculate anisotropy from relaxation times. Hiscott and Andrews [149] calculated parameters of the orbitals in cadmium from X-ray emission microscopy data. Kuznetsov et al. [150] calculated heats of melting using simplex. Pelletier et al. [151] performed calculations on the optical constants of thin films. Rousson et al. [152] calculated vibrational force constants. Castenmiller and Buck [153] used simplex for quantum mechanical calculations involving potential energy surfaces. DeCoen and Ralston [154] used simplex for conformational analysis. Mezey et al. [155] used it for calculations of transition states. Cheng and McCreery [156] performed theoretical calculations on chronoamperometry data using simplex, Muller and Brown [157] used the algorithm for theoretical studies on vinyl amines, and Gey and Kuhnel [158] used it for olefin calculations. Shiner et al. [159] used simplex to help calculate reaction mechanisms, Chestnut and Whitehurst [160] calculated theoretical carbon-13 NMR chemical shifts, and Wrobleski and Brown [161] calculated magnetic susceptibilities. Yoshida et al. [162] performed numerical analysis on steroid formation.

Simplex has been used on a number of occasions for function minimization [163–168]. It has also been used for optimizing experimental designs [169]. The field of pattern recognition has also seen the simplex algorithm applied by Ritter et al. [170], Wilkins [171], Takahashi et al. [172, 173], and Kaberline and Wilkins [174]–this last application using the supermodified simplex algorithm [175]. Specific applications of simplex-pattern recognition have included NMR spectra [176] and mass spectra [177].

REFERENCES

1. D. E. Long, "Simplex optimization of the response from chemical systems," *Anal. Chim. Acta*, **46**(2), 193–206 (1969).

2. M. J. Houle, D. E. Long, and D. Smette, "A simplex optimized colorimetric method for formaldehyde," *Anal. Lett.*, **3**(8), 401–408 (1970).

3. A. S. Olansky and S. N. Deming, "Optimization and interpretation of absorbance response in the determination of formaldehyde with chromotropic acid," *Anal. Chim. Acta*, **83**(1), 241–249 (1976).

4. F. P. Czech, "Simplex optimized J-acid method for the determination of formaldehyde," *J. Assoc. Offic. Anal. Chem.*, **56**(6), 1489–1495 (1973).

5. F. P. Czech, "Simplex optimized acetylacetone method for formaldehyde," *J. Assoc. Offic. Anal. Chem.*, **56**(6), 1496–1502 (1973).

6. S. N. Deming and P. G. King, "Computers and experimental optimization," *Research/Development*, **25**(5), 22–26 (1974).

7. S. N. Deming and S. L. Morgan, "Simplex optimization of variables in analytical chemistry," *Anal. Chem.*, **45**(3), 278A–283A (1973).
8. S. L. Morgan and S. N. Deming, "Simplex optimization of analytical chemical methods," *Anal. Chem.*, **46**(9), 1170–1181 (1974).
9. C. Vanroelen, R. Smits, P. Van den Windel, and D. L. Massart, "Application of factor analysis and simplex technique to the optimization of a phosphate determination via molybdenum blue," *Fresenius' Z. Anal. Chem.*, **280**(1), 21–23 (1976).
10. G. E. Mieling, R. W. Taylor, L. G. Hargis, J. English, and H. L. Pardue, "Fully automated stopped-flow studies with a hierarchical computer controlled system," *Anal. Chem.*, **48**, 1686–1693 (1976).
11. M. L. H. Turoff and S. N. Deming, "Optimization of the extraction of iron(II) from water into cyclohexane with hexafluoroacetylacetone and tri-*n*-butyl phosphate," *Talanta*, **24**(9), 567–571 (1977).
12. R. J. McDevitt and B. J. Barker, "Simplex optimization of the synergic extraction of a bis(diketo)copper(II) complex," *Anal. Chim. Acta*, **122**, 223–226 (1980).
13. C. L. Shavers, M. L. Parsons, and S. N. Deming, "Simplex optimization of chemical systems," *J. Chem. Educ.*, **56**, 307–309 (1979).
14. T. Michalowski, A. Rokosz, and E. Wojcik, "Optimization of the conventional method for determination of zinc as 8-hydroxyquinolate in alkaline tartrate medium," *Chem. Anal. (Warsaw)*, **25**, 563–566 (1980) [*Chem. Abstr.*, **94**(16), 131609u].
15. A. S. Olansky, L. R. Parker, Jr., S. L. Morgan, and S. N. Deming, "Automated development of analytical chemical methods. The determination of serum calcium by the cresolphthalein complexone method," *Anal. Chim. Acta*, **95**(3–4), 107–133 (1977).
16. G. Taguchi, *Introduction to Quality Engineering: Designing Quality into Products and Processes*, Kraus International Publications, White Plains, NY, 1986.
17. P. J. Ross, *Taguchi Techniques for Quality Engineering: Loss Function, Orthogonal Experiments, Parameter and Tolerance Design*, McGraw-Hill, New York, 1988.
18. A. Bendell, J. Disney, and W. A. Pridmore, Eds., *Taguchi Methods: Applications in World Industry*, IFS Publications, Springer-Verlag, London, 1989.
19. R. D. Krause and J. A. Lott, "Use of the simplex method to optimize analytical conditions in clinical chemistry," *Clin. Chem. (Winston-Salem, NC)*, **20**(7), 775–782 (1974).
20. J. A. Lott and K. Turner, "Evaluation of Trinder's glucose oxidase method for measuring glucose in serum and urine," *Clin. Chem. (Winston-Salem, NC)*, **21**(12), 1754–1760 (1975).
21. W. D. Basson, P. P. Pille, and A. L. Du Preez, "Automated *in situ* preparation of azomethine H and the subsequent determination of boron in aqueous solution," *The Analyst (London)*, **99**, 168–170 (1974).
22. A. S. Olansky and S. N. Deming, "Automated development of a kinetic method for the continuous-flow determination of creatinine," *Clin. Chem. (Winston-Salem, NC)*, **24**, 2115–2124 (1978).

23. H. Winicov, J. Schainbaum, J. Buckley, G. Longino, J. Hill, and C. E. Berkoff, "Chemical process optimization by computer – a self-directed chemical synthesis system," *Anal. Chim. Acta*, **103**, 469–476 (1978).

24. C. H. Lochmuller, K. R. Lung, and K. R. Cousins, "Applications of optimization strategies in the design of intelligent laboratory robotic procedures," *Anal. Lett.*, **18**(A4), 439–448 (1985).

25. R. R. Ernst, "Measurement and control of magnetic field homogeneity," *Rev. Sci. Instrum.*, **39**(7), 998–1012 (1968).

26. D. M. Cantor and J. Jonas, "Automated measurement of spin-lattice relaxation times: Optimized pulsed nuclear magnetic resonance spectrometry," *Anal. Chem.*, **48**, 1904–1906 (1976).

27. K. Hyakuna and G. L. Samuelson, "Shimplex – automatic control system for Y and C shims," *JEOL News*, **16A**(1), 17–19 (1980).

28. M. M. Siegel, "The use of the modified simplex method for automatic phase correction in Fourier-transform nuclear magnetic resonance spectroscopy," *Anal. Chim. Acta*, **133**, 103–108 (1981).

29. L. R. Parker, Jr., S. L. Morgan, and S. N. Deming, "Simplex optimization of experimental factors in atomic absorption spectrometry," *Appl. Spectrosc.*, **29**(5), 429–433 (1975).

30. E. R. Johnson, C. K. Mann, and T. J. Vickers, "Computer controlled system for study of pulsed hollow cathode lamps," *Appl. Spectrosc.*, **30**(4), 415–422 (1976).

31. R. G. Michel, J. M. Ottaway, J. Sneddon, and G. S. Fell, "Reproducible method for the preparation and operation of microwave-excited electrodeless discharge lamps for use in atomic-fluorescence spectrometry. Additional experience with cadmium lamps," *The Analyst (London)*, **103**, 1204–1209 (1978).

32. R. G. Michel, J. Coleman, and J. D. Winefordner, "A reproducible method for preparation and operation of microwave excited electrodeless discharge lamps: SIMPLEX optimization of experimental factors for a cadmium lamp," *Spectrochim. Acta*, Part B, **33B**(5), 195–215 (1978).

33. J. Borszeki, K. Doerffel, and E. Gegus, "Application of experimental planning methods in chemical research. III. Optimization of calcium atomic absorption determination using the simplex method," *Magy. Kem. Foly.*, **86** (1980) 207–210 [Hung., *Chem. Abstr.*, **93**(4), 36232m].

34. I. Taufer and J. Tauferova, "Experiment planning in determining the operational regime of an atomic absorption spectrophotometer by the simplex method," *Chem. Prum.*, **31** (1981) 16–20 (Czech.); [*Chem. Abstr.*, **94**, 95145d].

35. F. Steiglich, R. Stahlberg, W. Kluge, "Precision of flame-atomic-absorption-spectrometric determination of main components. 4. Optimization of the cobalt, copper, and chromium determination by applying statistical methods and the SIMPLEX method," *Fresenius' Z. Anal. Chem.*, **317**(5), 527–538 (1984) [Ger., *Chem. Abstr.*, **100**(22), 184984m].

36. R. L. Belchamber, D. Betteridge, A. P. Wade, A. J. Cruickshank, and P. Davison, "Removal of a matrix effect in ICP-AES multi-element analysis by simplex optimization," *Spectrochim. Acta (B)*, **41B**(5), 503–505 (1986).

37. L. Ebdon, M. R. Cave, and D. J. Mowthorpe, "Simplex optimization of inductively coupled plasmas," *Anal. Chim. Acta*, **115**, 179–187 (1980).

38. L. Ebdon, "The optimization of an inductively coupled plasma for metallurgical analysis," in R. M. Barnes, Ed., *Proceedings of the International Winter*

Conference on Developments in Atomic Plasma Spectrochemical Analysis 1980, Heyden, London, 1981, p. 94–110 [*Chem. Abstr.*, **96**, 173462].

39. M. R. Cave, D. M. Kaminaris, L. Ebdon, and D. J. Mowthorpe, "Fundamental studies of the application of an inductively coupled plasma to metallurgical analysis," *Anal. Proc. (London)*, **18**(1), 12–14 (1981).

40. S. P. Terblanche, K. Visser, and P. B. Zeeman, "The modified sequential simplex method of optimization as applied to an inductively coupled plasma source," *Spectrochim. Acta*, Part B, **36B**, 293–297 (1981).

41. J. J. Leary, A. E. Brooks, A. F. Dorrzapf, Jr., and D. W. Golightly, "An objective function for optimization techniques in simultaneous multiple-element analysis by inductively coupled plasma spectrometry," *Appl. Spectrosc.*, **36**, 37–40 (1982).

42. J. J. Brocas, "Simplex technique for nonlinear optimization: Its application to oil analysis by inductively coupled plasma emission spectrometry," *Analusis*, **1982**, 387–389 (1982).

43. J. Gonzalez, F. Berti, and J. L. Lamazares, "Simplex optimization of experimental factors in atomic absorption spectroscopy. Part 1," *Rev. Cubana Fis.*, **3**(2), 141–153 (1983) [*Chem. Abstr.*, **101**(12), 103052b].

44. J. Gonzalez, F. Berti, and M. Hernandez Martinez, "Simplex optimization of experimental factors in atomic spectroscopy. Part 2. Applications in the atomic emission spectroscopy," *Rev. Cubana Fis.*, **3**(3), 111–123 (1983).

45. L. R. Parker, Jr., N. H. Tioh, and R. M. Barnes, "Optimization approaches to the determination of arsenic and selenium by hydride generation and ICP-AES," *Appl. Spectrosc.*, **39**(1), 45–48 (1985).

46. L. R. Parker, Jr., M. R. Cave, and R. M. Barnes, "Comparison of simplex algorithms," *Anal. Chim. Acta*, **175**, 231–237 (1985).

47. G. R. Kornblum, J. Smeyers-Verbeke, Y. Michotte, A. Klok, D. L. Massart, and L. de Galan, "Optimization and long term stability of the induction coupled plasma in a synthetical biological matrix," in P. Braetter and P. Schramel, Eds., *Trace Element–Analytical Chemistry in Medicine and Biology*, Volume 2, Walter de Gruyter, Berlin, 1983, pp. 1161–1173.

48. M. R. Cave, R. M. Barnes, and P. Denzer, paper no. 24, *1982 Winter Conference on Plasma Spectroscopy*, Orlando, FL, 1982.

49. R. M. Barnes and P. Fodor, "Analysis of urine using inductively-coupled plasma emission spectroscopy with graphite rod electrothermal vaporization," *Spectrochim. Acta*, **38B**, 1191–1202 (1983).

50. R. C. Carpenter and L. Ebdon, "A comparison of inductively coupled plasma torch–sample introduction configurations using simplex optimization," *J. Anal. Atomic Spectrom.*, **1**(4), 265–268 (1986).

51. G. L. Moore, P. J. Humphries-Cuff, and A. E. Watson, "Simplex optimization of a nitrogen-cooled argon inductively coupled plasma for multielement analysis," *Spectrochim. Acta (B)*, **39B**, 915–929 (1984).

52. W. E. Rippetoe, E. R. Johnson, and T. J. Vickers, "Characterization of the plume of a direct current plasma arc for emission spectrometric analysis," *Anal. Chem.*, **47**, 436–440 (1975).

53. V. G. Belikov, V. E. Godyatskii, and A. I. Sichko, "Application of simplex planning of experiments to study the optimal conditions for analysis of chloracizin by differential photometry," *Farmatsiya (Moscow)*, **20**(3), 30–33 (1971) [Russ., *Chem. Abstr.*, **75**, 67539z].

54. I. Y. Kul and L. E. Kechatova, "Use of simplex experiment planning for selecting optimal conditions for the differential spectrophotometric determination of dimecarbine," *Farmatsiya (Moscow)*, **25**, 23–26 (1976) [Russ., *Chem. Abstr.*, **85**, 25446t].

55. B. B. Jablonski, W. Wegscheider, and D. E. Leyden, "Evaluation of computer directed optimization for energy dispersive x-ray spectrometry," *Anal. Chem.*, **51**, 2359–2364 (1979).

56. S. V. Lyubimova, S. V. Mamikonyan, Y. N. Svetailo, and K. I. Shchekin, "Optimization of the composition of substances for an experiment in calibration of a multielement radiometric x-ray analyzer," *Vopr. At. Nauki Tekh.*, *[Ser.]: Radiats. Tekh.*, **19**, 176–178 (1980) [Russ., *Chem. Abstr.*, **95**, 125405d].

57. A. Mangia, "A new approach to the problem of kaolinite interference in the determination of chrysolite asbestos by means of x-ray diffraction," *Anal. Chim. Acta*, **117**, 337–342 (1980).

58. S. Stieg and T. A. Nieman, "Application of a microcomputer controlled chemiluminescence research instrument to the simultaneous determination of cobalt(II) and silver(I) by gallic acid chemiluminescence," *Anal. Chem.*, **52**, 800–804 (1980).

59. C. Zimmermann and W. E. Hoehne, "Simplex optimization of a fluorometric determination of the pyruvate kinase and phosphofructoskinase activities from rabbit muscle using fluorescent adenine nucleotides," *Z. Med. Laboratoriumsdiagn.*, **21**, 259–267 (1980) [Ger., *Chem. Abstr.*, **94**(1), 1422s].

60. G. Wuensch, N. Czech, and G. Hegenberg, "Determination of tungsten with the capacitively coupled microwave plasma (CMP). Optimization of a CMP using factorial design and simplex method," *Z. Anal. Chem.*, **310**, 62–69 (1982) [Ger., *Chem. Abstr.*, **96**, 192515y].

61. H. A. Spaink, T. T. Lub, G. Kateman, and H. C. Smit, "Automation of the optical alignment of a diode-laser spectrometer by means of simplex optimization," *Anal. Chim. Acta*, **184**, 87–97 (1986).

62. L. de Galan and G. R. Kornblum, "Computer control of monochromators in atomic spectrometry," *Chem. Mag. (Rijswijk, Neth.)*, 240241 (1982) [*Chem. Abstr.*, **97**(22), 192324f].

63. D. Betteridge, T. J. Sly, A. P. Wade, and D. G. Porter, "Versatile automatic development system for flow injection analysis," *Anal. Chem.*, **58**(11), 2258–2265 (1986).

64. J. W. Elling, L. J. De Koning, F. A. Pinkse, N. M. M. Nibbering, M. M. Nico, and H. C. Smit, "Computer-controlled simplex optimization on a Fourier-transform ion cyclotron resonance mass spectrometer," *Anal. Chem.*, **61**(4), 330–334 (1989).

65. S. L. Morgan and S. N. Deming, "Optimization strategies for the development of gas-liquid chromatographic methods," *J. Chromatogr.*, **112**, 267–285 (1975).

66. S. L. Morgan and S. N. Deming, "Experimental optimization of chromatographic systems," *Sep. Purif. Methods*, **5**(2), 333–360 (1976).

67. J. Holderith, T. Toth, and A. Varadi, "Minimizing the time for gas chromatographic analysis. Search for optimal operational parameters by a simplex method," *J. Chromatogr.*, **119**, 215–222 (1976).

68. V. E. Medyantsev, D. A. Vyakhirev, and M. Y. Shtaerman, "Improvement in the gas chromatographic separation of resinates using a mathematical method of experiment planning," *Gidroliz. Lesokhim. Prom-st.*, **5**, 18–20 (1978) [Russ., *Chem. Abstr.*, **89**, 181417x].

69. S. L. Morgan and C. A. Jacques, "Response surface evaluation and optimization in gas chromatography," *J. Chromatogr. Sci.*, **16**, 500–505 (1978).

70. M. Singliar and L. Koudelka, "Optimization of the conditions of chromatographic analysis of technical mixtures of glycol ethers," *Chem. Prum.*, **29**, 134–139 (1979) [Slo., *Chem. Abstr.*, **91**(2), 13182q].

71. O. V. Sakhartova, V. Sates, J. Freimanis, and A. Avots, "Chromatography of prostaglandins, their analogs and precursors. I. Gas chromatographic control of the stages of 2-(6-carboethoxyhexyl)-6-endo-vinylbicyclo[3.1=0]hexan-1-one synthesis," *Latv. PSR Zinat. Akad. Vestis, Kim Ser.*, **4**, 414–420 (1981) [Russ., *Chem. Abstr.*, **95**, 180312w].

72. F. J. Yang, A. C. Brown III, and S. P. Cram, "Splitless sampling for capillary-column gas chromatography," *J. Chromatogr.*, **158**, 91–109 (1978).

73. F. Hsu, J. Anderson, and A. Zlatkis, "A practical approach to optimization of a selective gas-chromatographic detector by a sequential simplex method," *J. High Res. Chrom. Chrom. Commun.*, **3**, 648–650 (1980).

74. V. Svoboda, "Search for optimal eluent composition for isocratic liquid column chromatography," *J. Chromatogr.*, **201**, 241–252 (1980).

75. M. L. Rainey and W. C. Purdy, "Simplex optimization of the separation of phospholipids by high-pressure liquid chromatography," *Anal. Chim. Acta*, **93**, 211–219 (1977).

76. M. W. Watson and P. W. Carr, "Simplex algorithm for the optimization of gradient elution high-performance liquid chromatography," *Anal. Chem.*, **51**, 1835–1842 (1979).

77. J. C. Berridge, "Optimization of high-performance liquid chromatographic separations with the aid of a microcomputer," *Anal. Proc. (London)*, **19**(10), 472–475 (1982).

78. J. C. Berridge, "Unattended optimisation of reversed-phase high-performance liquid chromatographic separations using the modified simplex algorithm," *J. Chromatogr.*, **244**, 1–14 (1982).

79. J. C. Berridge, "Unattended optimization of normal phase high-performance liquid chromatography separations with a microcomputer controlled chromatograph," *Chromatographia*, **16**, 172–174 (1982).

80. J. C. Berridge and E. G. Morrisey, "Automated optimization of reversed-phase high-performance liquid chromatography separations," *J. Chromatogr.*, **316**, 69–79 (1984).

81. X. Tomas, J. Hernandez, and L. G. Sabate, "The use of the simplex method in the optimization of chromatographic separations," *Afinidad*, **36**, 485–488 (1979) [Span., *Chem. Abstr.*, **93**(4), 32165u].

82. F. H. Walters and S. N. Deming, "Window diagrams versus the sequential simplex method: Which is correct?," *Anal. Chim. Acta*, **167**, 361–363 (1985).

83. J. L. Glajch and L. R. Snyder, Eds., *Computer-Assisted Method Development for High-Performance Liquid Chromatography*, Elsevier, Amsterdam, 1990.

84. L. P. Kofman, V. G. Gorskii, B. Z. Brodskii, A. A. Sergo, T. P. Nozdrina, A. I. Osipov, and Y. V. Nazarov, "Use of optimization methods in developing a process of obtaining prometrine," *Zavod. Lab.*, **34**, 69–71 (1968).

85. W. K. Dean, K. J. Heald, and S. N. Deming, "Simplex optimization of reaction yields," *Science*, **189**, 805–806 (1975).

86. G. Dumenil, A. Cremieux, R. Phan Tan Luu, and J. P. Aune, "Bioconversion from DL-homoserine to L-threonine. II. Application of the simplex method of optimization," *Eur. J. Appl. Microbiol.*, **1**, 221–231 (1975).

87. P. W. Dillon and G. R. Underwood, "Cyclic allenes. 2. The conversion of cyclo-propylidenes to allenes. A simplex-INDO study," *J. Am. Chem. Soc.*, **99**(8), 2435–2446 (1977).

88. R. D. Gilliom, W. P. Purcell, and T. R. Bosin, "Sequential simplex optimization applied to drug design in the indole, 1-methylindole, and benzo[b]thiophene series," *Eur. J. Med. Chem.-Chim. Ther.*, **12**(2), 187–192 (1977).

89. R. Lazaro, P. Bouchet, and R. Jacquier, "Experimental design. II. Simplex optimization of the synthesis of an industrial pyrazolone," *Bull. Soc. Chim. Fr.*, **11–12**, pt. 2, 1171–1174 (1977) [Fr., *Chem. Abstr.*, **89**(5), 42106g].

90. R. Lazaro, D. Mathieu, R. Phan Tan Luu, and J. Elguero, "Experimental design. I. Analysis of the formation mechanism of 1,3- and 1,5-dimethylpy-razoles by action of methylhydrazine on 4,4-dimethoxy-2-butanone in acid medium," *Bull. Soc. Chim. Fr.*, **11–12**, pt. 2, 1163–1170 (1977) [Fr., *Chem. Abstr.*, **89**(1), 5761w].

91. D. S. Amenta, C. E. Lamb, and J. J. Leary, "Simplex optimization of yield of *sec*-butylbenzene in a Friedel–Crafts alkylation. A special undergraduate project in analytical chemistry," *J. Chem. Educ.*, **56**, 557–558 (1979).

92. D. Brunel, J. Itier, A. Commeyras, R. Phan Tan Luu, and D. Mathieu, "Les acides perfluorosulfoniques. II. Activation du n-pentane par les systemes superacides du type $RFSO_3H$-SbF_5. Recherche des conditions optimales dans le cas des acides $C_4F_3SO_3H$ et CF_3SO_3H," *Bull. Soc. Chim. Fr.*, **5**, 257–263 (1978) (Fr.).

93. E. Li-Chan, N. Helbig, E. Holbek, S. Chau, and S. Nakai, "Covalent attachment of lysine to wheat gluten for nutritional improvement," *J. Agric. Food. Chem.*, **27**(4), 877–882 (1979).

94. F. L. Chubb, J. T. Edward, and S. C. Wong, "Simplex optimization of yields in the Bucherer–Bergs reaction," *J. Org. Chem.*, **45**, 2315–2320 (1980).

95. G. Minkova and V. Baeva, "Optimization of the isomerization of *p*-(isoamy-loxy) anilinium thiocyanate to [*p*-(isoamyloxy)phenyl] thiourea. I. Experimental optimization by the simplex method," *Khim. Ind. (Sofia)*, **9**, 402–404 (1980) [Bulg., *Chem. Abstr.*, **95**, 24485z].

96. F. Darvas, "Application of the sequential simplex method in designing drug analogs," *J. Med. Chem.*, **17**(8), 799–804 (1974).

97. F. Darvas, L. Kovacs, and A. Eory, "Computer optimization by the sequential simplex method in designing drug analogs," *Abh. Akad. Wiss. DDR, Abt. Math., Naturwiss., Tech.*, (2N, Quant. Struct.-Act. Anal.), 311–315 (1978) [*Chem. Abstr.*, **91**(5), 32476e].

✓ 98. W. D. Baasel, "Exploring response surfaces to establish optimum conditions," *Chem. Eng. (NY)*, **72**(22), 147–152 (1965).

√ 99. B. H. Carpenter and H. C. Sweeney, "Process improvement with 'simplex' self-directing evolutionary operation," *Chem. Eng. (NY)*, **72**(14), 117–126 (1965).

100. I. C. Kenworthy, "Some examples of simplex evolutionary operation in the paper industry," *Appl. Stat.*, **16**, 211–224 (1967).

101. A. Avots, V. Ulaste, and G. Enins, "Optimization of a heterogeneous catalytic process by the method of simplex planning," *Latv. PSR Zinat. Akad. Vestis, Kim. Ser.*, **6**, 717–721 (1970) [Russ., *Chem. Abstr.*, **74**, 55689q].

102. A. G. Antonenkov and V. M. Pomerantsev, "Use of a simplex method to search for the optimum composition of a catalyst," *Zh. Prikl. Khim. (Leningrad)*, **47**, 899–900 (1974) [Russ., *Chem. Abstr.*, **81**, 30064p].

103. R. Tymczynski, Z. Spych, and W. Kupsc, "Sequential application of the simplex method for optimization of rubber-blend composition," *Polimery*, 15, 530–532 (1970) [Pol., *Chem. Abstr.*, 74, 127123a].

104. H. Brusset, D. Depeyre, M. Boeda, R. Melkior, and J. L. Staedtsbaeder, "Optimization of a multistage reactor for oxidation of sulfur dioxide by three direct methods. (Hooke-Jeeves, Rosenbrock, and simplex)," *Can. J. Chem. Eng.*, 49(6), 786–791 (1971) [Fr., *Chem. Abstr.*, 76(12), 61372p].

105. R. F. Reiss and A. J. Katz, "Optimizing recovery of platelets in platelet rich plasma by the simplex strategy," *Transfusion*, 16, 370–374 (1976).

106. G. L. Silver, "Simplex characterization of equilibrium: Application to plutonium," *Radiochem. Radioanal. Lett.*, 27(4), 243–248 (1976).

107. J. Ceynowa and R. Wodzki, "Simplex optimization of carbon electrodes for the hydrogen-oxygen membrane fuel cell," *J. Power Sources*, 1(4), 323–331 (1977).

108. W. J. Evans, V. L. Frampton, and A. D. French, "A comparative analysis of the interaction of mannitol with borate by calorimetric and pH techniques," *J. Phys. Chem.*, 81, 1810–1812 (1977).

109. M. G. Davydov and A. P. Naumov, "Optimization of multielement quantitative activation analysis," *Radiochem. Radioanal. Lett.*, 35, 77–84 (1978).

110. D. D. Burgess, "Optimization of multielement instrumental neutron activation analysis," *Anal. Chem.*, 57, 1433–1436 (1985).

111. D. D. Burgess and P. Hayumbu, "Simplex optimization by advance prediction for single-element instrumental neutron activation analysis," *Anal. Chem.*, 56, 1440–1443 (1984).

112. M. K. Hanafey, R. L. Scott, T. H. Ridgway, and C. N. Reilley, "Analysis of electrochemical mechanisms by finite difference simulation and simplex fitting of double potential step current, charge, and absorbance response," *Anal. Chem.*, 50, 116–137 (1978).

113. F. Vlacil and Huynh Dang Khanh, "Determination of low concentrations of dibenzyl sulfoxide in aqueous solutions," *Collect. Czech. Chem. Commun.*, 44(6) 1908–1917 (1979) [*Chem. Abstr.*, 91(24), 203879v].

114. S. Bolanca and A. Golubovic, "The determination of ink composition by means of the simplex method," *Hem. Ind.*, 34, 168–172 (1980) [Serbo-Croatian, *Chem. Abstr.*, 93, 96911s].

115. J. R. Chipperfield, A. C. Hayter, and D. E. Webster, "Reactivity of main-group-transition-metal bonds. Part II. The kinetics of reaction between tricarbonyl (eta-cyclopentadienyl)(trimethylstannyl)-chromium and iodine," *J. Chem. Soc., Dalton Trans.*, 20, 2048–2050 (1975).

116. C. L. McMinn and J. H. Ottaway, "Studies on the mechanism and kinetics of the 2-oxyglutarate dehydrogenase system from pig heart," *Biochem. J.*, 161, 569–581 (1977).

117. R. J. Matthews, S. R. Goode, and S. L. Morgan, "Characterization of an enzymatic determination of arsenic using response surface methodology," *Anal. Chim. Acta*, 133, 169–182 (1981).

118. M. Otto and G. Werner, "Optimization of a kinetic-catalytic method by the use of a numerical model and the simplex method," *Anal. Chim. Acta*, 128, 177–183 (1981).

119. F. Mayne, "Optimization techniques and galenic formulation: Example of sequential simplex and compressed tablets," *Expo-Congr. Int. Technol. Pharm.*, 1st, 5, 65–84 (1977) (Fr.), Assoc. Pharm. Galenique Ind.: Chatenay-Malabry, Fr. [*Chem. Abstr.*, 90(6), 43769h].

120. E. Shek, M. Ghani, and R. E. Jones, "Simplex search in optimization of capsule formulation," *J. Pharm. Sci.*, **69**, 1135–1142 (1980).

121. G. Kostal, G. B. Ashe, and M. P. Eastman, "Generation of component set representations for rocks using the simplex approach," *J. Chem. Geol.*, **39**(4), 347–356 (1983).

122. J. A. Nelder and R. Mead, "A simplex method for function minimization," *Comput. J.*, **7**, 308–313 (1965).

123. W. Spendley, G. R. Hext, and F. R. Himsworth, "Sequential application of simplex designs in optimisation and evolutionary operation," *Technometrics*, **4**(4), 441–461 (1962).

124. W. E. Wentworth, "Rigorous least squares adjustment: Application to some non-linear equations," *J. Chem. Educ.*, **42**, 96–103, 162–167 (1965).

125. N. R. Draper and H. Smith, *Applied Regression Analysis*, 2nd ed., Wiley, New York, 1981.

126. R. O'Neill, "Function minimization using a simplex procedure," *Appl. Stat.*, **20**, 338–345 (1971).

127. M. S. Caceci and W. P. Cacheris, "Fitting curves to data," *BYTE*, 340–362 (May, 1984).

128. W. H. Press, B. P. Flannery, S. A. Teukolsky, and W. T. Vetterline, *Numerical Recipes in Pascal: The Art of Scientific Computing*, Cambridge University Press, Cambridge, 1989.

129. P. C. Jurs, *Computer Software Applications in Chemistry*, John Wiley, New York, 1986, Chapter 9, pp. 125–140.

130. M. F. Delaney and D. M. Mauro, "Extension of multicomponent self-modeling curve resolution based on a library of reference spectra," *Anal. Chim. Acta*, **172**, 193 (1985).

131. D. M. Mauro and M. F. Delaney, "Resolution of infrared spectra of mixtures by self-modelling curve resolution using a library of reference spectra with simplex-assisted searching," *Anal. Chem.*, **58**(13), 2622–2628 (1986).

132. C. Dauwe, M. Dorikens, and L. Dorikens-Vanpraet, "Analysis of double decay spectra by the simplex stepping method," *Appl. Phys.*, **5**, 45–47 (1974).

133. J. W. Akitt, "Visual and automatic spectral analysis using a small digital computer," *Appl. Spectrosc.*, **29**, 493–496 (1975).

134. R. H. Geiss and T. C. Huang, "Quantitative x-ray energy dispersive analysis with the transition electron microscope," *X-Ray Spectrom.*, **4**(4), 196–201 (1975).

135. B. Moraweck, P. deMontgolfier, and A. J. Renouprez, "X-ray line-profile analysis. I. A method of unfolding diffraction profiles," *J. Appl. Crystallogr.*, **10**, 184–190 (1977).

136. C. E. Fiori and R. L. Myklbust, "A simplex method for fitting Gaussian profiles to x-ray spectra obtained with an energy-dispersive detector," *DOE Symp. Ser.* 1978, 49 (Comput. Act. Anal. Gamma-Ray Spectrosc.) 139–149 (1979) [*Chem. Abstr.*, **92**, 157089h].

137. C. F. Lam, A. Forst, and H. Bank, "Simplex: a method for spectral deconvolution applicable to energy dispersion analysis," *Appl. Spectrosc.*, **33**, 273–278 (1979).

138. S. K. Silber, R. A. Deans, and R. A. Geanangel, "A comparison of the simplex and Gauss iterative algorithms for curve fitting in Moessbauer spectra," *Comput. Chem.*, **4**, 123–130 (1980).

139. H. N. Cheng, "Markovian statistics and simplex algorithm for carbon-13 nuclear magnetic resonance spectra of ethylene-propylene copolymers," *Anal. Chem.*, **54**, 1828–1833 (1982).

140. D. J. Leggett, "Numerical analysis of multicomponent spectra," *Anal. Chem.*, **49**(2), 276–281 (1977).

141. B. W. Madsen and J. S. Robertson, "Improved parameter estimates in drug-protein binding studies by nonlinear regression," *J. Pharm. Pharmacol.*, **26**, 807–813 (1974).

142. Y. C. Martin and J. J. Hackbarth, "Examples of the application of non-linear regression analysis to chemical data," in B. R. Kowalski, Ed., *Chemometrics: Theory and Application*, ACS Symposium Series 52, American Chemical Society, Washington, D.C., 1977, Chapter 8, pp. 153–164.

143. F. R. Van de Voort, C-Y. Ma, and S. Nakai, "Molecular weight distribution of interacting proteins calculated by multiple regression analysis from sedimentation equilibrium data: An interpretation of α_{s1}-j-casein interaction," *Arch. Biochem. Biophys.*, **195**, 596–606 (1979).

144. P. V. Pederson, "Curve fitting and modeling in pharmacokinetics and some practical experiences with NONLIN and a new program FUNFIT," *J. Pharmacokinet. Biopharmaceut.*, **5**, 512–531 (1977).

145. M. C. Kohn, L. E. Menten, and D. Garfinkel, "A convenient computer program for fitting enzymatic rate laws to steady-state data." *Comput. Biomed. Res.*, **12**, 461–469 (1979).

146. G. Just, U. Lindner, W. Pritzkow, and M. Roellig, "Diels-Alder reactions. V. Kinetic modelling of reactions occurring during the codimerization of cyclopentadiene with 1,3-butadiene," *J. Prakt. Chem.*, **317**(6), 979–989 (1975) [Ger., *Chem. Abstr.*, **84**, 58166j].

147. S. Becker, H. J. Kohler, and C. Weiss, "Geometric optimization of small molecules in the all valence electron-MO formulism using the SIMPLEX and gradient methods," *Collect. Czech. Chem. Commun.*, **40**(3), 794–798 (1975) [Ger., *Chem. Abstr.*, **83**, 65820g].

148. S. Berger, F. R. Kreissl, D. M. Grant, and J. D. Roberts, "Determination of anisotropy of molecular motion with 13-C spin-lattice relaxation times," *J. Am. Chem. Soc.*, **97**, 1805–1808 (1975).

149. L. A. Hiscott and P. T. Andrews, "Cd 4d spin-orbit splittings in alloys of Cd and Mg," *J. Phys., F, Metal Phys.*, **5**, 1077–1082 (1975).

150. G. M. Kuznetsov, M. P. Leonor, S. K. Kuznetsova, and V. I. Kovalev, "Calculation of the heat of melting of compounds and of simple substances," *Zh. Fiz. Khim.*, **50**, 2517–2521 (1976) [*Chem. Abstr.*, **86**, 22607f].

151. E. Pelletier, P. Roche, and B. Vidal, "Automatic evaluation of optical constants and thickness of thin films: Application to thin dielectric layers," *Nouv. Rev. Opt.*, **7** (1976) 353–362 [Fr., *Chem. Abstr.*, **86**, 81012f].

152. R. Rousson, G. Tantot, and M. Tournarie, "Numerical determination of vibrational force constants by means of a simplex minimizing method," *J. Mol. Spectrosc.*, **59**(1), 1–7 (1976).

153. W. A. M. Castenmiller and H. M. Buck, "A quantum chemical study of the $C_6H_5^+$ potential energy surface. Evidence for a nonclassical pyramidal carbenic species," *Recl. Trav. Chim. Pays-Bas*, **96**, 2017–213 (1977) [*Chem. Abstr.*, **87**, 183780t].

154. J.-L. DeCoen and E. Ralston, "Theoretical conformational analysis of Asn^1, Val^5 angiotensin II," *Biopolymers*, **16**, 1929–1943 (1977).

155. P. G. Mezey, M. R. Peterson, and I. G. Csizmadia, "Transition state determination by the X-method," *Can. J. Chem.*, **55**(16), 2941–2945 (1977).

156. H-Y. Cheng and R. L. McCreery, "Simultaneous determination of reversible potential and rate constant for a first-order EC reaction by potential dependent chronoamperometry," *Anal. Chem.*, **50**(4), 645–648 (1978).

157. K. Muller and L. D. Brown, "Enamines. 1. Vinyl amine – theoretical study of its structure, electrostatic potential, and proton affinity," *Helv. Chim. Acta*, **61**, 1407–1418 (1978).

158. E. Gey and W. Kuhnel, "Berechnung von potentialflachenausschnitten radikalischer anlagerungsreaktionen an olefinische doppelbindungen semiempirische untersuchungene der reaktion ·CH_3 + C_2H_4," *Collect. Czech. Chem. Commun.*, **44**, 3649–3655 (1979) (Ger.).

159. V. J. Shiner, Jr., D. A. Nollen, and K. Humski, "Multiparameter optimization procedure for the analysis of reaction mechanistic schemes. Solvolyses of cyclopentyl p-bromobenzenesulfonate," *J. Org. Chem.*, **44**(13), 2108–2115 (1979).

160. D. B. Chestnut and F. W. Whitehurst, "A simplex optimized INDO calculation of ^{13}C chemical shifts in hydrocarbons," *J. Comput. Chem.*, **1**, 36–45 (1980).

161. J. T. Wrobleski and D. B. Brown, "A study of the variable-temperature magnetic susceptibility of two Ti(III) oxalate complexes," *Inorg. Chem. Acta*, **38**, 227–230 (1980).

162. T. Yoshida, M. Sueki, H. Taguchi, S. Kulprecha, and N. Nilubol, "Modelling and optimization of steroid transformation in a mixed culture," *Eur. J. Appl. Microbiol. Biotechnol.*, **11**, 81–88 (1981).

163. D. M. Olsson, "A sequential simplex program for solving minimization problems," *J. Qual. Tech.*, **6**, 53–57 (1974).

164. R. O'Neill, "Corrigendum. Function minimization using a simplex procedure," *Appl. Stat.*, **23**, 252 (1974).

165. F. James and M. Roos, "MINUIT– a system for function minimization and analysis of the parameter errors and correlations," *Comp. Phys. Commun.*, **10**, 343–367 (1975).

166. P. R. Benyon, "Remark: Function minimization using a simplex procedure," *Appl. Stat.*, **25**, 97 (1976).

167. J. W. Akitt, "Function minimization using the Nelder and Mead simplex method with limited arithmetic precision: The self regenerative simplex," *Comput. J.*, **20**, 84–85 (1977).

168. K. Mueller and L. D. Brown, "Location of saddle points and minimum energy paths by a constrained simplex optimization procedure," *Theor. Chim. Acta*, **53**(1), 75–93 (1979).

169. J. W. Evans, "Computer augmentation of experimental designs to maximize [X'X]," *Technometrics*, **21**, 321–330 (1979).

170. G. L. Ritter, S. R. Lowry, C. L. Wilkins, and T. L. Isenhour, "Simplex pattern recognition," *Anal. Chem.*, **47**(12), 1951–1956 (1975).

171. C. L. Wilkins, "Interactive pattern recognition in the chemical analysis laboratory," *J. Chem. Inf. Comput. Sci.*, **17**(4), 242–249 (1977).

172. Y. Takahashi, Y. Miyashita, H. Abe, and S. Sasaki, "A new approach for ordered multicategorical classification using simplex technique," *Brunseki Kagaku*, **33**(11), E487–E494 (1984).

173. Y. Takahashi, Y. Miyashita, Y. Tanaka, H. Hayasaka, H. Abe, and S. Sasaki, "Discriminative structural analysis using pattern recognition techniques

in the structure-taste problem of perillartines," *J. Pharm. Sci.*, **73**(6), 737–741 (1984).

174. S. L. Kaberline and C. L. Wilkins, "Evaluation of the super-modified simplex for use in chemical pattern recognition," *Anal. Chim. Acta*, **103**, 417–428 (1978).

175. M. W. Routh, P. A. Swartz, and M. B. Denton, "Performance of the super modified simplex," *Anal. Chem.*, **49**(9), 1422–1428 (1977).

176. T. R. Brunner, C. L. Wilkins, T. F. Lam, L. J. Soltzberg, and S. L. Kaberline, "Simplex pattern recognition applied to carbon-13 nuclear magnetic resonance spectrometry," *Anal. Chem.*, **48**(8), 1146–1150 (1976).

177. T. F. Lam, C. L. Wilkins, T. R. Brunner, L. J. Soltzberg, and S. L. Kaberline, "Large-scale mass spectral analysis by simplex pattern recognition," *Anal. Chem.*, **48**(12), 1768–1774 (1976).

Chapter 11

Simplex Optimization Bibliography

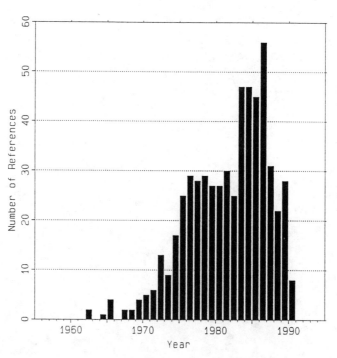

Figure 11.1 Number of sequential simplex references vs. year.

1962:

F. R. Himsworth, "Empirical methods of optimisation," *Trans. Inst. Chem. Eng.*, **40**, 345–349 (1962).

W. Spendley, G. R. Hext, and F. R. Himsworth, "Sequential application of simplex designs in optimisation and evolutionary operation," *Technometrics*, **4**(4), 441–461 (1962).

1964:

C. W. Lowe, "Some techniques of evolutionary operation," *Trans. Inst. Chem. Eng.*, **42**, T334–T344 (1964).

1965:

W. D. Baasel, "Exploring response surfaces to establish optimum conditions," *Chem. Eng. (NY)*, **72**(22), 147–152 (1965).

M. J. Box, "A new method of constrained optimization and a comparison with other methods," *Comput. J.*, **8**, 42–52 (1965).

B. H. Carpenter and H. C. Sweeney, "Process improvement with 'simplex' self-directing evolutionary operation," *Chem. Eng. (NY)*, **72**(14), 117–126 (1965).

J. A. Nelder and R. Mead, "A simplex method for function minimization," *Comput. J.*, **7**, 308–313 (1965).

1967:

I. C. Kenworthy, "Some examples of simplex evolutionary operation in the paper industry," *Appl. Stat.*, **16**, 211–224 (1967).

C. W. Lowe, "A report on a simplex evolutionary operation for multiple responses," *Trans. Inst. Chem. Eng.*, **45**(1), T3–T7 (1967).

1968:

R. R. Ernst, "Measurement and control of magnetic field homogeneity," *Rev. Sci. Instrum.*, **39**(7), 998–1012 (1968).

L. P. Kofman, V. G. Gorskii, B. Z. Brodskii, A. A. Sergo, T. P. Nozdrina, A. I. Osipov, and Y. V. Nazarov, "Use of optimization methods in developing a process of obtaining prometrine," *Zavod. Lab.*, **34**, 69–71 (1968).

1969:

D. E. Long, "Simplex optimization of the response from chemical systems," *Anal. Chim. Acta*, **46**(2), 193–206 (1969).

B. K. Selinger and R. Speed, "Simple unit for studying transient fluorescent species in the nanosecond range," *Chem. Instrum.*, **2**(1), 91–102 (1969).

W. Spendley, "Nonlinear least squares fitting using a modified simplex minimization method," in R. Fletcher, Ed., *Optimization and its Applications*, Symposium of the Institute of Mathematics, University of Keele, England, 1968, Academic Press, New York, 1969, pp. 259–270.

T. Umeda, "Optimal design of an absorber-stripper system," *Ind. Eng. Chem. Process Des. Dev.*, **8**(3), 308–317 (1969).

1970:

A. Avots, V. Ulaste, and G. Enins, "Optimization of a heterogeneous catalytic process by the method of simplex planning," *Latv. PSR Zinat. Akad. Vestis, Kim. Ser.*, **6**, 717–721 (1970) [Russ., *Chem. Abstr.*, **74**, 55689q].

G. S. G. Beveridge and R. S. Schecter, *Optimization: Theory and Practice*, McGraw-Hill, New York, 1970, pp. 367–385.

R. M. Driver, "Statistical methods and the chemist," *Chem. Br.*, **6**, 154–158 (1970).

D. M. Himmelblau, *Process Analysis by Statistical Methods*, John Wiley, New York, 1970, pp. 181–186.

M. J. Houle, D. E. Long, and D. Smette, "A simplex optimized colorimetric method for formaldehyde," *Anal. Lett.*, **3**(8), 401–408 (1970).

R. Tymczynski, Z. Spych, and W. Kupsc, "Sequential application of the simplex method for optimization of rubber-blend composition," *Polimery (Warsaw)*, **15**, 530–532 (1970) [Pol., *Chem. Abstr.*, **74**, 127123a].

1971:

V. G. Belikov, V. E. Godyatskii, and A. I. Sichko, "Application of simplex planning of experiments to study the optimal conditions for analysis of chloracizin by differential photometry," *Farmatsiya (Moscow)*, **20**(3), 30–33 (1971) [Russ., *Chem. Abstr.*, **75**, 67539z].

H. Brusset, D. Depeyre, M. Boeda, R. Melkior, and J. L. Staedtsbaeder, "Optimization of a multistage reactor for oxidation of sulfur dioxide by three direct methods. (Hooke-Jeeves, Rosenbrock, and simplex)," *Can. J. Chem. Eng.*, **49**(6), 786–791 (1971) [Fr., *Chem. Abstr.*, **76**(12), 61372p].

S. S. Lee, L. E. Erickson, and L. T. Fan, "Modelling and optimization of a tower-type activated sludge system," in R. P. Canale, Ed., *Biological Waste Treatment*, Interscience, New York, 1971, pp. 141–173 [*Chem. Abstr.*, **75**(24), 143760g].

R. O'Neill, "Function minimization using a simplex procedure," *Appl. Stat.*, **20**, 338–345 (1971).

V. V. Pomazkov and A. V. Krylova, "Determination of the composition of complex additives to cements by the successive simplex planning method," *Izv. Vyssh. Uchebn. Zaved., Stroit. Arkhitekt.*, **14**(6), 78–81 (1971) [Russ., *Chem. Abstr.*, **75**(24), 143624r].

T. Umeda and A. Ichikawa, "A modified complex method for optimization," *Ind. Eng. Chem. Process Des. Dev.*, **10**(2), 229–236 (1971).

1972:

V. V. Abakumov, A. I. Golomzik, V. A. Kryuchkov, and N. P. Vedernikova, "Comparison of factor and simplex planning methods (for leaching copper from ores from the Blyavinsk deposit)," *Tr. Ural. Nauch.-Issled. Proekt. Inst. Medn. Prom.*, No. 15, 177–179 (1972) [Russ., *Chem. Abstr.*, **80**(16), 85924d].

R. H. Davis and J. H. Ottaway, "Application of optimization procedures to tracer kinetic data," *Math. Biosci.*, **13**(3–4), 265–282 (1972).

M. J. S. Dewar and P. Weiner, "Ground states of molecules. XXIII. MINDO/2 calculations for naphthalene," *Theor. Chim. Acta*, **27**(4), 373–375 (1972).

U. V. Ganin, R. V. Kolesnikova, R. M. Besisskii, I. E. Latyshev, and I. I. Litvinenko, "Design calculation of a regulator for automatic temperature control in an oxidizing column by experimental-statistical methods on an analog computer," *Automat. Kontr.-Izmar. Prib., Nauch.-Tekh. Sb.*, No. 12, 16–19 (from: *Ref. Zh., Khim.*, 1973, Abstr. No. 2I244) (1972) [Russ., *Chem. Abstr.*, **80**(18), 97911k].

D. M. Himmelblau, *Applied Nonlinear Programming*, McGraw-Hill, New York, 1972.

D. M. Himmelblau, "A uniform evaluation of unconstrained optimization techniques," in F. A. Lootsma, Ed., *Numerical Methods for Non-Linear Optimization*, Academic Press, New York, 1972, Chapter 6, pp. 69–97.

A. Parczewski and A. Rokosz, "Optimization of catalysts preparation process. I. Theoretical approach," *Chem. Stosow.*, **16**(4), 409–422 (1972) [Pol., *Chem. Abstr.*, **78**(20), 128831z].

A. Parczewski, A. Rokosz, and P. Laidler, "Optimization of catalysts preparation process. II. Experimental studies," *Chem. Stosow.*, **16**(4), 423–433 (1972) [Pol., *Chem. Abstr.*, **78**(20), 128832a].

J. M. Parkinson and D. Hutchinson, "A consideration of non-gradient algorithms for the unconstrained optimization of functions of high dimensionality," in F. A. Lootsma, Ed., *Numerical Methods for Non-Linear Optimization*, Academic Press, New York, 1972, Chapter 7, pp. 99–113.

J. M. Parkinson and D. Hutchinson, "An investigation into the efficiency of variants on the simplex method," in F. A. Lootsma, Ed., *Numerical Methods for Non-Linear Optimization*, Academic Press, New York, 1972, Chapter 8, pp. 115–135.

D. A. Phillips, "A preliminary investigation of function optimisation by a combination of methods," *Comput. J.*, **17**, 75–79 (1972).

T. Umeda, A. Shindo, and E. Tazaki, "Optimal design of chemical process by a feasible decomposition method," *Ind. Eng. Chem. Process Des. Dev.*, **11**, 1–8 (1972).

J. D. Wren and R. C. Harshman, "Effect of radiant energy transfer and chlorophyll concentration on the growth kinetics of *Chlorella pyrenoidosa* TX71105 in a laminar flow photosynthetic gas exchanger during low intensity illumination," *AIChE Symp. Ser.*, **68**(124), 238–249 (1972) [*Chem. Abstr.*, **77**(19), 123907y].

1973:

G. R. Atwood and W. W. Foster, "Transformation of bounded variables in simplex optimization techniques," *Ind. Eng. Chem. Process Des. Dev.*, **12**(4), 485–486 (1973).

W. E. Biles, "An accelerated sequential simplex search technique," *AIIE Trans.*, **5**, 127–134 (1973).

F. P. Czech, "Simplex optimized J-acid method for the determination of formalde-hyde," *J.-Assoc. Off. Anal. Chem.*, **56**(6), 1489–1495 (1973).

F. P. Czech, "Simplex optimized acetylacetone method for formaldehyde," *J.-Assoc. Off. Anal. Chem.*, **56**(6), 1496–1502 (1973).

S. N. Deming and S. L. Morgan, "Simplex optimization of variables in analytical chemistry," *Anal. Chem.*, **45**(3), 278A–283A (1973).

R. W. Glass and D. F. Bruley, "REFLEX method for empirical optimization," *Ind. Eng. Chem. Process Des. Dev.*, **12**(1), 6–10 (1973).

D. L. Keefer, "Simpat: Self-bounding direct search method for optimization," *Ind. Eng. Chem. Process Des. Dev.*, **12**(1), 92–99 (1973).

J. L. Kuester and J. H. Mize, *Optimization Techniques with FORTRAN*, McGraw-Hill, New York, 1973, pp. 298–308.

L. A. Tysall, "Application of mathematical designs to experimental work in the coat-ings field," *Farbe Lack*, **79**(7), 622–633 (1973) [Ger., *Chem. Abstr.*, **79**(12), 67865k].

1974:

P. R. Adby and M. A. H. Dempster, *Introduction to Optimization Methods*, Chapman and Hall, London, 1974, pp. 45–48.

J. Aerts, A. Gawdzik, W. Krajweski, and J. Skrzypek, "Identification of parameters in reaction rate equations for complex chemical reaction systems," *Pr. Nauk. Inst. Inz. Chem. Urzadzen Cieplnych Politech. Wroclaw.*, **24**, 71–86 (1974) [Pol., *Chem. Abstr.*, **82**(23), 155671m].

A. G. Antonenkov and V. M. Pomerantsev, "Use of a simplex method to search for the optimum composition of a catalyst," *Zh. Prikl. Khim. (Leningrad)*, **47**, 899–900 (1974) [Russ., *Chem. Abstr.*, **81**, 30064p].

W. D. Basson, P. P. Pille, and A. L. Du Preez, "Automated *in situ* preparation of azomethine H and the subsequent determination of boron in aqueous solution," *Analyst (London)*, **99**, 168–170 (1974).

J. M. Chambers and J. R. Ertel, "A remark on algorithm AS 47 'Function minimiza-tion using a simplex procedure'," *Appl. Stat.*, **23**, 250–251 (1974).

F. Darvas, "Application of the sequential simplex method in designing drug ana-logs," *J. Med. Chem.*, **17**(8), 799–804 (1974).

C. Dauwe, M. Dorikens, and L. Dorikens-Vanpraet, "Analysis of double decay spec-tra by the simplex stepping method," *Appl. Phys.*, **5**, 45–47 (1974).

S. N. Deming and P. G. King, "Computers and experimental optimization," *Research/Development*, **25**(5), 22–26 (1974).

G. J. Hahn and A. F. Dershowitz, "Evolutionary operation to-day–some survey results and observations," *Appl. Stat.*, **23**, 214–218 (1974).

P. G. King and S. N. Deming, "UNIPLEX: Single-factor optimization of response in the presence of error," *Anal. Chem.*, **46**(11), 1476–1481 (1974).

R. D. Krause and J. A. Lott, "Use of the simplex method to optimize analytical con-ditions in clinical chemistry," *Clin. Chem. (Winston-Salem, NC)*, **20**(7), 775–782 (1974).

J. M. Leaman, "A simple way to optimize," *Machine Design*, **46**, 204–208 (1974).

C. W. Lowe, "Evolutionary operation in action," *Appl. Stat.*, **23**, 218–226 (1974).

B. W. Madsen and J. S. Robertson, "Improved parameter estimates in drug-protein binding studies by nonlinear regression," *J. Pharm. Pharmacol.*, **26**, 807–813 (1974).

S. L. Morgan and S. N. Deming, "Simplex optimization of analytical chemical methods," *Anal. Chem.*, **46**(9), 1170–1181 (1974).

D. M. Olsson, "A sequential simplex program for solving minimization problems," *J. Qual. Tech.*, **6**, 53–57 (1974).

R. O'Neill, "Corrigendum. Function minimization using a simplex procedure," *Appl. Stat.*, **23**, 252 (1974).

L. A. Yarbro and S. N. Deming, "Selection and preprocessing of factors for simplex optimization," *Anal. Chim. Acta*, **73**(2), 391–398 (1974).

M. M. Yusipov and Ch. N. Yusupbekova, "Modelling and optimizing ion-exchange synthesis of thallium salts," *Tr. Tasshk. Politekh. Inst.*, **119**, 58–60 (1974) [Russ., *Chem. Abstr.*, **84**(16), 107824h].

1975:

J. W. Akitt, "Visual and automatic spectral analysis using a small digital computer," *Appl. Spectrosc.*, **29**, 493–496 (1975).

K. A. Andrianov, B. A. Kamaritskii, B. I. D'yachenko, and V. D. Nedorosol, "Concerning effective experimentation during minimization of the consumption of raw material components," *Dokl. Akad. Nauk SSSR*, **225**(5), 1135–1138 (1975) [Russ., *Chem. Abstr.*, **84**(16), 107681j].

S. Becker, H. J. Kohler, and C. Weiss, "Geometric optimization of small molecules in the all valence electron-MO formulism using the SIMPLEX and gradient methods," *Collect. Czech. Chem. Commun.*, **40**(3), 794–798 (1975) [Ger., *Chem. Abstr.*, **83**, 65820g].

S. Berger, F. R. Kreissl, D. M. Grant, and J. D. Roberts, "Determination of anisotropy of molecular motion with 13-C spin-lattice relaxation times," *J. Am. Chem. Soc.*, **97**, 1805–1808 (1975).

J. R. Chipperfield, A. C. Hayter, and D. E. Webster, "Reactivity of main-group-transition-metal bonds. Part II. The kinetics of reaction between tricarbonyl (eta-cyclopentadienyl)(trimethylstannyl)-chromium and iodine," *J. Chem. Soc., Dalton Trans.*, **20**, 2048–2050 (1975).

L. L. Combs and M. Holloman, "Semiempirical calculations of internal barriers to rotation and ring puckering. I. Present techniques," *J. Phys. Chem.*, **79**(5), 512–521 (1975).

W. K. Dean, K. J. Heald, and S. N. Deming, "Simplex optimization of reaction yields," *Science*, **189**, 805–806 (1975).

G. Dumenil, A. Cremieux, R. Phan Tan Luu, and J. P. Aune, "Bioconversion from DL-homoserine to L-threonine. II. Application of the simplex method of optimization," *Eur. J. Appl. Microbiol.*, **1**, 221–231 (1975).

R. H. Geiss and T. C. Huang, "Quantitative x-ray energy dispersive analysis with the transition electron microscope," *X-Ray Spectrom.*, **4**(4), 196–201 (1975).

L. A. Hisscott and P. T. Andrews, "Cd 4d spin-orbit splittings in alloys of Cd and Mg," *J. Phys. F: Met. Phys.*, **5**, 1077–1082 (1975).

J. Holderith, T. Toth, and A. Varadi, "Minimizing the time for gas chromatographic analysis. Finding optimal measuring parameters by means of a simplex method," *Magy. Kem. Foly.*, **81**(4), 162–164 (1975) [Hung., *Chem. Abstr.*, **83**(8), 71167j].

F. James and M. Roos, "MINUIT– a system for function minimization and analysis of the parameter errors and correlations," *Comp. Phys. Commun.*, **10**, 343–367 (1975).

G. Just, U. Lindner, W. Pritzkow, and M. Roellig, "Diels–Alder reactions. V. Kinetic modelling of reactions occurring during the codimerization of cyclopentadiene with 1,3-butadiene," *J. Prakt. Chem.*, **317**(6), 979–989 (1975) [Ger., *Chem. Abstr.*, **84**, 58166j].

P. G. King, S. N. Deming, and S. L. Morgan, "Difficulties in the application of simplex optimization to analytical chemistry," *Anal. Lett.*, **8**(5), 369–376 (1975).

J. A. Lott and K. Turner, "Evaluation of Trinder's glucose oxidase method for measuring glucose in serum and urine," *Clin. Chem. (Winston-Salem, NC)*, **21**(12), 1754–1760 (1975).

M. Meus, A. Parczewski, and A. Rokosz, "Optimization of analytical methods. Determination of zinc by direct titration with potassium," *Chem. Anal. (Warsaw)*, **20**(2), 247–256 (1975) [*Chem. Abstr.*, **83**(14), 125663p].

S. L. Morgan and S. N. Deming, "Optimization strategies for the development of gas-liquid chromatographic methods," *J. Chromatogr.*, **112**, 267–285 (1975).

D. M. Olsson and L. S. Nelson, "The Nelder–Mead simplex procedure for function minimization," *Technometrics*, **17**, 45–51 (1975).

L. R. Parker, Jr., S. L. Morgan, and S. N. Deming, "Simplex optimization of experimental factors in atomic absorption spectrometry," *Appl. Spectrosc.*, **29**(5), 429–433 (1975).

W. E. Rippetoe, E. R. Johnson, and T. J. Vickers, "Characterization of the plume of a direct current plasma arc for emission spectrometric analysis," *Anal. Chem.*, **47**, 436–440 (1975).

G. L. Ritter, S. R. Lowry, C. L. Wilkins, and T. L. Isenhour, "Simplex pattern recognition," *Anal. Chem.*, **47**(12), 1951–1956 (1975).

R. Smits, C. Vanroelen, and D. L. Massart, "The optimization of information obtained by multicomponent chromatographic separation using the simplex technique," *Fresenius' Z. Anal. Chem.*, **273**(1), 1–5 (1975).

J. Subert and J. Cizmarik, "Simplex method for empirical optimizing and its use in resolving problems of pharmaceutical analysis," *Pharmazie*, **30**(12), 761–765 (1975) [Ger., *Chem. Abstr.*, **84**(18), 126790k].

G. R. Walsh, *Methods of Optimization*, John Wiley, London, 1975, pp. 81–84.

H. Walter, Jr. and D. Flannigan, "Detection of atmospheric pollutants. Correlation technique," *Appl. Opt.*, **14**(6), 1423–1428 (1975).

1976:

M. Auriel, *Nonlinear Programming*, Prentice-Hall, Englewood Cliffs, NJ, 1976, pp. 245–247.

P. R. Benyon, "Remark: function minimization using a simplex procedure," *Appl. Stat.*, **25**, 97 (1976).

D. Boris, "Use of statistical methods in experimental studies in the process industry. IV. How to use EVOP for optimizing a technological process," *Hem. Ind.*, **30**(8), 467–470 (1976) [Serbo-Croation, *Chem. Abstr.*, **86**(10), 57429b].

T. R. Brunner, C. L. Wilkins, T. F. Lam, L. J. Soltzberg, and S. L. Kaberline, "Simplex pattern recognition applied to carbon-13 nuclear magnetic resonance spectrometry," *Anal. Chem.*, **48**(8), 1146–1150 (1976).

D. M. Cantor and J. Jonas, "Automated measurement of spin-lattice relaxation times: Optimized pulsed nuclear magnetic resonance spectrometry," *Anal. Chem.*, **48**, 1904–1906 (1976).

J. A. Clements and L. F. Prescott, "Data point weighting in pharmacokinetic analysis. Intravenous paracetamol in man," *J. Pharm. Pharmacol.*, **28**(9), 707–709 (1976).

S. N. Deming, S. L. Morgan, and M. R. Willcott, "Sequential simplex optimization," *Am. Lab.*, **8**(10), 13–19 (1976).

M. R. Detaevernier, L. Dryon, and D. L. Massart, "The separation of some tricyclic antidepressants by high-performance liquid chromatography," *J. Chromatogr.*, **128**, 204–207 (1976).

T. J. Dols and B. H. Armbrecht, "Simplex optimization as a step in method development," *J.-Assoc. Off. Anal. Chem.*, **59**(6), 1204–1207 (1976).

M. J. Faddy, "A note on the general time-dependent stochastic compartmental model," *Biometrics*, **32**, 443–448 (1976).

D. J. Finney, "Radioligand assay," *Biometrics*, **32**, 721–740 (1976).

N. M. Fisher and R. M. Dvorkina, "Development of a solution for cold phosphate coating of nickel using a simplex optimization method," *Issled. Obl. Khim. Istochnikov Toka*, **4**, 152–156 (1976) [*Chem. Abstr.*, **88**(24), 175278r].

G. J. Hahn, "Process improvement through simplex EVOP," *CHEMTECH*, **6**(5), 343–345 (1976).

J. Holderith, T. Toth, and A. Varadi, "Minimizing the time for gas chromatographic analysis. Search for optimal operational parameters by a simplex method," *J. Chromatogr.*, **119**, 215–222 (1976).

E. R. Johnson, C. K. Mann, and T. J. Vickers, "Computer controlled system for study of pulsed hollow cathode lamps," *Appl. Spectrosc.*, **30**(4), 415–422 (1976).

I. Y. Kul and L. E. Kechatova, "Use of simplex experiment planning for selecting optimal conditions for the differential spectrophotometric determination of dimecarbine," *Farmatsiya (Moscow)*, **25**, 23–26 (1976) [Russ., *Chem. Abstr.*, **85**, 25446t].

G. M. Kuznetsov, M. P. Leonor, S. K. Kuznetsova, and V. I. Kovalev, "Calculation of the heat of melting of compounds and of simple substances," *Zh. Fiz. Khim.*, **50**, 2517–2521 (1976) [*Chem. Abstr.*, **86**, 22607f].

T. F. Lam, C. L. Wilkins, T. R. Brunner, L. J. Soltzberg, and S. L. Kaberline, "Large-scale mass spectral analysis by simplex pattern recognition," *Anal. Chem.*, **48**(12), 1768–1774 (1976).

G. E. Mieling, R. W. Taylor, L. G. Hargis, J. English, and H. L. Pardue, "Fully automated stopped-flow studies with a hierarchical computer controlled system," *Anal. Chem.*, **48**, 1686–1693 (1976).

S. L. Morgan and S. N. Deming, "Experimental optimization of chromatographic systems," *Sep. Purif. Methods*, **5**(2), 333–360 (1976).

T. Nakagawa and T. Nagai, "Interaction between serum albumin and mercaptoun-decahydrododecaborate ion (an agent for boron-neutron capture therapy of brain tumors). II. Proposal of two interaction models," *Chem. Pharm. Bull.*, **24**(12), 2942–2948 (1976) [*Chem. Abstr.*, **86**(15), 101138g].

A. S. Olansky and S. N. Deming, "Optimization and interpretation of absorbance response in the determination of formaldehyde with chromotropic acid," *Anal. Chim. Acta*, **83**(1), 241–249 (1976).

E. Pelletier, P. Roche, and B. Vidal, "Automatic evaluation of optical constants and thickness of thin films: Application to thin dielectric layers," *Nouv. Rev. Opt.*, **7**, 353–362 (1976) [Fr., *Chem. Abstr.*, **86**, 81012f].

D. B. Popovic and P. Mihajlovic, "Optical properties of gallium indium arsenide (GaxIn1-xAs) in infrared spectra," *Fizika (Zagreb)*, **8**, Suppl., 79–81 (1976) [*Chem. Abstr.*, **87**(12), 93027y].

R. F. Reiss and A. J. Katz, "Optimizing recovery of platelets in platelet rich plasma by the simplex strategy," *Transfusion*, **16**, 370–374 (1976).

R. Rousson, G. Tantot, and M. Tournarie, "Numerical determination of vibrational force constants by means of a simplex minimizing method," *J. Mol. Spectrosc.*, **59**(1), 1–7 (1976).

G. L. Silver, "Simplex characterization of equilibrium: application to plutonium," *Radiochem. Radioanal. Lett.*, **27**(4), 243–248 (1976).

C. Vanroelen, R. Smits, P. Van den Winkel, and D. L. Massart, "Application of factor analysis and simplex technique to the optimization of a phosphate determination via molybdenum blue," *Fresenius' Z. Anal. Chem.*, **280**(1), 21–23 (1976).

R. Wodzki and J. Ceynowa, "Simplex design method for planning optimum experiments," *Wiad. Chem.*, **30**(5), 337–347 (1976) [Pol., *Chem. Abstr.*, **86**, 31497x].

1977:

J. W. Akitt, "Function minimization using the Nelder and Mead simplex method with limited arithmetic precision: the self-regenerative simplex," *Comput. J.*, **20**, 84–85 (1977).

V. A. Burtsev, A. A. Kondakov, A. M. Timonin, and V. F. Shanskii, "Optimization of pulsed carbon dioxide laser parameters," *Kvantaovaya Electron. (Moscow)*, **4**(11), 2374–2378 (1977) [Russ., *Chem. Abstr.*, **88**(18), 128788x].

W. A. M. Castenmiller and H. M. Buck, "A quantum chemical study of the $C_6H_5^+$ potential energy surface. Evidence for a nonclassical pyramidal carbenic species," *Recl. Trav. Chim. Pays-Bas*, **96**, 2017–213 (1977) [*Chem. Abstr.*, **87**, 183780t].

J. Ceynowa and R. Wodzki, "Simplex optimization of carbon electrodes for the hydrogen-oxygen membrane fuel cell," *J. Power Sources*, **1**(4), 323–331 (1977).

J.-L. DeCoen and E. Ralston, "Theoretical conformational analysis of Asn^1, Val^5 angiotensin II," *Biopolymers*, **16**, 1929–1943 (1977).

S. N. Deming, "Optimization of experimental parameters in chemical analysis," *ACS Symp. Ser.*, **63** (Validation Meas. Process, Symp.), 162–175 (1977) [*Chem. Abstr.*, **88**(6), 44503n].

S. N. Deming and S. L. Morgan, "Advances in the application of optimization methodology in chemistry," in B. R. Kowalski, Ed., *Chemometrics: Theory and Application*, ACS Symposium Series 52, American Chemical Society, Washington, D.C., 1977, Chapter 1, pp. 1–13.

P. W. Dillon and G. R. Underwood, "Cyclic allenes. 2. The conversion of cyclopropylidenes to allenes. A simplex-INDO study," *J. Am. Chem. Soc.*, **99**(8), 2435–2446 (1977).

W. J. Evans, V. L. Frampton, and A. D. French, "A comparative analysis of the interaction of mannitol with borate by calorimetric and pH techniques," *J. Phys. Chem.*, **81**, 1810–1812 (1977).

R. D. Gilliom, W. P. Purcell, and T. R. Bosin, "Sequential simplex optimization applied to drug design in the indole, 1-methylindole, and benzo[b]thiophene series," *Eur. J. Med. Chem. Chim. Ther.*, **12**(2), 187–192 (1977).

P. Gravereau, A. Hardy, and A. Bonnin, "MIFe(CrO$_4$)$_2$ series: crystal structure of alpha-ammonium iron chromate NH$_4$Fe(CrO$_4$)$_2$," *Acta Crystallogr., Sect. B: Struct. Sci.*, **B33**(5), 1362–1367 (1977).

K. Hayakawa and J. Okada, "Gas phase reaction of trimethylenediamine and propylene glycol in the reactor packed with alumina catalyst – search for the optimum reaction conditions of homopiperazine yield by experimental design," *Yakugaku Zasshi*, **97**(12), 1299–1304 (1977) [Japan., *Chem. Abstr.*, **89**(3), 24267f].

R. Lazaro, P. Bouchet, and R. Jacquier, "Experimental design. II. Simplex optimization of the synthesis of an industrial pyrazolone," *Bull. Soc. Chim. Fr.*, **11–12**, pt. 2, 1171–1174 (1977) [Fr., *Chem. Abstr.*, **89**(5), 42106g].

R. Lazaro, D. Mathieu, R. Phan Tan Luu, and J. Elguero, "Experimental design. I. Analysis of the formation mechanism of 1,3- and 1,5-dimethylpyrazoles by action of methylhydrazine on 4,4-dimethoxy-2-butanone in acid medium," *Bull. Soc. Chim. Fr.*, **11–12**, pt. 2, 1163–1170 (1977) [Fr., *Chem. Abstr.*, **89**(1), 5761w].

D. J. Leggett, "Numerical analysis of multicomponent spectra," *Anal. Chem.*, **49**(2), 276–281 (1977).

Y. C. Martin and J. J. Hackbarth, "Examples of the application of non-linear regression analysis to chemical data," in B. R. Kowalski, Ed., *Chemometrics: Theory and Application*, ACS Symposium Series 52, American Chemical Society, Washington, D.C., 1977, Chapter 8, pp. 153–164.

F. Mayne, "Optimization techniques and galenic formulation: Example of sequential simplex and compressed tablets," *Expo-Congr. Int. Technol. Pharm., 1st*, **5**, 65–84 (1977) (Fr.), Assoc. Pharm. Galenique Ind.: Chatenay-Malabry, Fr. [*Chem. Abstr.*, **90**(6), 43769h].

C. L. McMinn and J. H. Ottaway, "Studies on the mechanism and kinetics of the 2-oxyglutarate dehydrogenase system from pig heart," *Biochem. J.*, **161**, 569–581 (1977).

P. G. Mezey, M. R. Peterson, and I. G. Csizmadia, "Transition state determination by the X-method," *Can. J. Chem.*, **55**(16), 2941–2945 (1977).

B. Moraweck, P. deMontgolfier, and A. J. Renouprez, "X-ray line-profile analysis. I. A method of unfolding diffraction profiles," *J. Appl. Crystallogr.*, **10**, 184–190 (1977).

A. S. Olansky, L. R. Parker, Jr., S. L. Morgan, and S. N. Deming, "Automated development of analytical chemical methods. The determination of serum calcium by the cresolphthalein complexone method," *Anal. Chim. Acta*, **95**(3–4), 107–133 (1977).

P. V. Pederson, "Curve fitting and modeling in pharmacokinetics and some practical experiences with NONLIN and a new program FUNFIT," *J. Pharmacokinet. Biopharmaceut.*, **5**, 512–531 (1977).

M. R. Peterson and I. G. Csizmadia, "Saddle points by the X-method," *Prog. Theor. Org. Chem.*, **2** (Appl. MO Theory Org. Chem.), 117–126 (1977) [*Chem. Abstr.*, **88**(21), 151864c].

M. L. Rainey and W. C. Purdy, "Simplex optimization of the separation of phospholipids by high-pressure liquid chromatography," *Anal. Chim. Acta*, **93**, 211–219 (1977).

M. W. Routh, P. A. Swartz, and M. B. Denton, "Performance of the super modified simplex," *Anal. Chem.*, **49**(9), 1422–1428 (1977).

M. L. H. Turoff and S. N. Deming, "Optimization of the extraction of iron(II) from water into cyclohexane with hexafluoroacetylacetone and tri-n-butyl phosphate," *Talanta*, **24**(9), 567–571 (1977).

L. L. Vidrevich, N. S. Afonskaya, and L. B. Chernyak, "Optimization of tanning using experimental-statistical methods," *Kozh.-Obuvn. Prom-st.*, **19**(6), 30–32 (1977) [Russ., *Chem. Abstr.*, **87**(10), 69768v].

C. L. Wilkins, "Interactive pattern recognition in the chemical analysis laboratory," *J. Chem. Inf. Comput. Sci.*, **17**(4), 242–249 (1977).

1978:

H-Y. Cheng and R. L. McCreery, "Simultaneous determination of reversible potential and rate constant for a first-order EC reaction by potential dependent chronoamperometry," *Anal. Chem.*, **50**(4), 645–648 (1978).

T-S. Chou, and L. L. Hegedus, "Transient diffusivity measurements in catalyst pellets with two zones of differing diffusivities," *AIChE J.*, **24**(2), 255–260 (1978).

F. Darvas, L. Kovacs, and A. Eory, "Computer optimization by the sequential simplex method in designing drug analogs," *Abh. Akad. Wiss. DDR, Abt. Math., Naturwiss., Tech.* (2N, Quant. Struct.-Act. Anal.), 311–315 (1978) [*Chem. Abstr.*, **91**(5), 32476e].

M. G. Davydov and A. P. Naumov, "Optimization of multielement quantitative activation analysis," *Radiochem. Radioanal. Lett.*, **35**, 77–84 (1978).

S. N. Deming, "Optimization of Methods," Chapter 2 in R. F. Hirsch, Ed., *Proceedings of the Eastern Analytical Symposium on Principles of Experimentation and Data Analysis*, Franklin Institute Press, Philadelphia, PA, 1978, pp. 31–55.

S. N. Deming and L. R. Parker, Jr., "A review of simplex optimization in analytical chemistry," *CRC Crit. Rev. Anal. Chem.*, **7**(3), 187–202 (1978).

J. W. Evans and A. R. Manson, "Optimal experimental designs in two dimensions using minimum bias estimation," *J. Am. Stat. Assoc.*, **73**, 171–176 (1978).

J. Gala and A. Budniok, "Mathematical methods for experimental planning and selection of optimal conditions for processes," *Pr. Nauk. Uniw. Slask. Katowicach*, **209**, 18–44 (1978) [Pol., *Chem. Abstr.*, **90**(4), 31002q].

M. K. Hanafey, R. L. Scott, T. H. Ridgway, and C. N. Reilley, "Analysis of electrochemical mechanisms by finite difference simulation and simplex fitting of double potential step current, charge, and absorbance response," *Anal. Chem.*, **50**, 116–137 (1978).

R. E. Hayes, J. I. Wadsworth, and J. J. Spadaro, "Corn- and wheat-based blended food formulations with cottonseed or peanut flour," *Cereal Foods World*, **23**(9), 548–553, 556 (1978).

S. L. Kaberline and C. L. Wilkins, "Evaluation of the super-modified simplex for use in chemical pattern recognition," *Anal. Chim. Acta*, **103**, 417–428 (1978).

G. Kleeman and K. Hartmann, "Optimization of petroleum processing units—contribution to process intensification," *Wiss. Z. Tech. Hochsch. "Carl Schorlemmer" Leuna-Merseberg*, **20**(4), 417–425 (1978) [Ger., *Chem. Abstr.*, **90**(16), 12424g].

T. Kobayashi, K. Tani, and S. Tamura, "Determination of optimum positions of the poloidal field coils of a large tokamak," *Jpn. J. Appl. Phys.*, **17**(12), 2139–2146 (1978).

R. J. Laplante, R. J. Thibert, and H. S. Asselstine, "An improved kinetic determination for creatinine using the Abbott ABA-100," *Microchem. J.*, **23**(4), 541–551 (1978).

C. H. Lin and S. C. Liu, "A new numerical method for automated spectral isolation of component substances in a set of mixtures," *J. Chin. Chem. Soc.*, **25**, 167–177 (1978).

D. L. Massart, A. Dijkstra, and L. Kaufman, *Evaluation and Optimization of Laboratory Methods and Analytical Procedures*, Elsevier, Amsterdam, 1978.

V. E. Medyantsev, D. A. Vyakhirev, and M. Y. Shtaerman, "Improvement in the gas chromatographic separation of resinates using a mathematical method of experiment planning," *Gidroliz. Lesokhim. Prom-st.*, **5**, 18–20 (1978) [Russ., *Chem. Abstr.*, **89**, 181417x].

A. Meglhorn, "MO-LCAO calculations on polymethines. VII. Effect of hydration and protonization on the electronic structure and the energy of simple stretopolymethines," *J. Signalaufzeichnungsmaterial*, **6**(3), 211–219 (1978) [Ger., *Chem. Abstr.*, **90**(11), 86496j].

R. G. Michel, J. M. Ottaway, J. Sneddon, and G. S. Fell, "Reproducible method for the preparation and operation of microwave-excited electrodeless discharge lamps for use in atomic-fluorescence spectrometry. Additional experience with cadmium lamps," *Analyst (London)*, **103**, 1204–1209 (1978).

R. G. Michel, J. Coleman, and J. D. Winefordner, "A reproducible method for preparation and operation of microwave excited electrodeless discharge lamps: SIMPLEX optimization of experimental factors for a cadmium lamp," *Spectrochim. Acta, Part B*, **33B**(5), 195–215 (1978).

S. L. Morgan and C. A. Jacques, "Response surface evaluation and optimization in gas chromatography," *J. Chromatogr. Sci.*, **16**, 500–505 (1978).

K. Muller and L. D. Brown, "Enamines. 1. Vinyl amine—theoretical study of its structure, electrostatic potential, and proton affinity," *Helv. Chim. Acta*, **61**, 1407–1418 (1978).

A. S. Olansky and S. N. Deming, "Automated development of a kinetic method for the continuous-flow determination of creatinine," *Clin. Chem. (Winston-Salem, NC)*, **24**, 2115–2124 (1978).

S. S. Rao, *Optimization: Theory and Applications*, Wiley Eastern Limited, New Delhi, 1978, pp. 284–292.

M. Suchanek, L. Sucha, and Z. Urner, "Optimization of analytical procedures," *Chem. Listy*, **72**, 1037–1042 (1978) [Czech., *Chem. Abstr.*, **90**, 33385r].

T. Takamatsu and Y. Shimizu, "An effective computational method of the optimal discrete control action of chemical process based on sensitivity analysis," *J. Chem. Eng. Jpn.*, **11**(3), 221–226 (1978) [*Chem. Abstr.*, **89**(10), 76981d].

F. Vlacil and V. Selepova, "Direct photometric determination of cobalt, copper, or iron in an organic phase containing sulfoxide," *Sb. Vys. Sk. Chem.-Technol. Praze, Anal. Chem.*, **H-13**, 119–129 (1978) [*Chem. Abstr.*, **93**(10), 106351a].

H. Winicov, J. Schainbaum, J. Buckley, G. Longino, J. Hill, and C. E. Berkoff, "Chemical process optimization by computer – a self-directed chemical synthesis system," *Anal. Chim. Acta*, **103**, 469–476 (1978).

F. J. Yang, A. C. Brown III, and S. P. Cram, "Splitless sampling for capillary-column gas chromatography," *J. Chromatogr.*, **158**, 91–109 (1978).

1979:

D. S. Amenta, C. E. Lamb, and J. J. Leary, "Simplex optimization of yield of *sec*-butylbenzene in a Friedel–Crafts alkylation. A special undergraduate project in analytical chemistry," *J. Chem. Educ.*, **56**, 557–558 (1979).

W. E. Biles and J. J. Swain, "Mathematical programming and the optimization of computer simulations," *Math. Program. Stud.*, **11**, 189–207 (1979).

G. F. Brissey, R. B. Spencer, and C. L. Wilkins, "High-speed algorithm for simplex calculations," *Anal. Chem.*, **51**, 2295–2297 (1979).

D. Brunel, J. Itier, A. Commeyras, R. Phan Tan Luu, and D. Mathieu, "Les acides perfluorosulfoniques. II. Activation du *n*-pentane par les systemes superacides du type $RFSO_3H$-SbF_5. Recherche des conditions optimales dans le cas des acides $C_4F_3SO_3H$ et CF_3SO_3H," *Bull. Soc. Chim. Fr.*, **5**, 257–263 (1978) (Fr.).

K. Doerffel and G. Ehrlich, "Process optimization and interpretation of measurements – necessities of modern analytical chemistry," *Wiss. Z.-Karl-Marx-Univ. Leipzig, Math.-Naturwiss. Reihe*, **28**, 459–465 (1979) [Ger., *Chem. Abstr.*, **92**, 68915g].

J. W. Evans, "Computer augmentation of experimental designs to maximize [X'X]," *Technometrics*, **21**, 321–330 (1979).

C. E. Fiori and R. L. Myklbust, "A simplex method for fitting Gaussian profiles to x-ray spectra obtained with an energy-dispersive detector," *DOE Symp. Ser.* 1978, 49 (Comput. Act. Anal. Gamma-Ray Spectrosc.) 139–149 (1979) [*Chem. Abstr.*, **92**, 157089h].

E. Gey and W. Kuhnel, "Berechnung von potentialflachenausschnitten radikalischer anlagerungsreaktionen an olefinische doppelbindungen semiempirische untersuchungene der reaktion $\cdot CH_3 + C_2H_4$," *Collect. Czech. Chem. Commun.*, **44**, 3649–3655 (1979) (Ger.).

P. Halmos, B. Pinter, and J. Inczedy, "Application of optimization methods in chemical research. I. Optimization of the binding of aspartic acid by ion exchange," *Magy. Kem. Foly.*, **85**(8), 341–344 (1979) [Hung., *Chem. Abstr.*, **92**(1), 6890m].

Huynh Dang Khanh and F. Vlacil, "Direct photometric determination of nickel in the organic phase after the extraction with dibenzyl sulfoxide," *Sb. Vys. Sk. Chem.-Technol. Praze, Anal. Chem.*, **H-14**, 191–196 (1979) [*Chem. Abstr.*, **94**(8), 57495k].

B. B. Jablonski, W. Wegscheider, and D. E. Leyden, "Evaluation of computer directed optimization for energy dispersive x-ray spectrometry," *Anal. Chem.*, **51**, 2359–2364 (1979).

M. C. Kohn, L. E. Menten, and D. Garfinkel, "A convenient computer program for fitting enzymatic rate laws to steady-state data," *Comput. Biomed. Res.*, **12**, 461–469 (1979).

C. F. Lam, A. Forst, and H. Bank, "Simplex: A method for spectral deconvolution applicable to energy dispersion analysis," *Appl. Spectrosc.*, **33**, 273–278 (1979).

E. Li-Chan, N. Helbig, E. Holbek, S. Chau, and S. Nakai, "Covalent attachment of lysine to wheat gluten for nutritional improvement," *J. Agric. Food Chem.*, **27**(4), 877–882 (1979).

R. G. Michel, J. M. Ottaway, J. Sneddon, and G. S. Fell, "Preparation and operation of selenium electrodeless discharge lamps for use in atomic-fluorescence flame spectroscopy," *Analyst (London)*, **104**, 687–691 (1979).

K. Mueller and L. D. Brown, "Location of saddle points and minimum energy paths by a constrained simplex optimization procedure," *Theor. Chim. Acta*, **53**(1), 75–93 (1979).

S. Nakai and F. Van de Voort, "Application of multiple regression analysis to sedimentation equilibrium data of α_{s1}- and F-casein interactions for calculation of molecular weight distributions," *J. Dairy Res.*, **46**(2), 283–290 (1979).

I. B. Rubin and C. K. Bayne, "Statistical designs for the optimization of the nitrogen-phosphorus gas chromatographic detector response," *Anal. Chem.*, **51**(4), 541–546 (1979).

C. L. Shavers, M. L. Parsons, and S. N. Deming, "Simplex optimization of chemical systems," *J. Chem. Educ.*, **56**, 307–309 (1979).

V. J. Shiner, Jr., D. A. Nollen, and K. Humski, "Multiparameter optimization procedure for the analysis of reaction mechanistic schemes. Solvolyses of cyclopentyl *p*-bromobenzenesulfonate," *J. Org. Chem.*, **44**(13), 2108–2115 (1979).

M. Singliar and L. Koudelka, "Optimization of the conditions of chromatographic analysis of technical mixtures of glycol ethers," *Chem. Prum.*, **29**, 134–139 (1979) [Slo., *Chem. Abstr.*, **91**(2), 13182q].

X. Tomas, J. Hernandez, and L. G. Sabate, "The use of the simplex method in the optimization of chromatographic separations," *Afinidad*, **36**, 485–488 (1979) [Span., *Chem. Abstr.*, **93**(4), 32165u].

Z. Urner, L. Sucha, and Z. Kohoutek, "Optimization of the spectrophotometric determination of ammonia by the indophenol method," *Sb. Vys. Sk. Chem.-Technol. Praze, Anal. Chem.*, **H-14**, 85–89 (1979) [Ger., *Chem. Abstr.*, **94**(8), 57494j].

F. R. Van de Voort, C-Y. Ma, and S. Nakai, "Molecular weight distribution of interacting proteins calculated by multiple regression analysis from sedimentation equilibrium data: An interpretation of α_{s1}-*k*-casein interaction," *Arch. Biochem. Biophys.*, **195**, 596–606 (1979).

F. Vlacil and Huynh Dang Khanh, "Determination of low concentrations of dibenzyl sulfoxide in aqueous solutions," *Collect. Czech. Chem. Commun.*, **44**(6) 1908–1917 (1979) [*Chem. Abstr.*, **91**(24), 203879v].

A. P. Wade, "Computer assisted optimization of chemical systems, in particular flow injection analysis," *Anal. Proc. (London)*, **20**(3), 108–110 (1979).

M. W. Watson and P. W. Carr, "Simplex algorithm for the optimization of gradient elution high-performance liquid chromatography," *Anal. Chem.*, **51**, 1835–1842 (1979).

W. Wegscheider, B. B. Jablonski, and D. E. Leyden, "Automated determination of optimum excitation conditions for single and multielement analysis with energy dispersive x-ray fluorescence spectrometry," *Adv. X-Ray Anal.*, **22**, 433–451 (1979).

1980:

H. L. Bank, A. J. Forst, and C. F. Lam, "QUEST: Quantitation using an elemental standardless technique for energy dispersion analysis," *Ultramicroscopy*, **5**(2), 153–162 (1980).

W. E. Biles and J. J. Swain, *Optimization and Industrial Experimentation*, John Wiley, New York, 1980.

H. Boennemann, W. Brijoux, and K. H. Simmrock, "Experimental optimization of organic syntheses. Homogeneous transition metal catalysis using statistical methods," *Erdoel Kohle, Erdgas, Petrochem.*, **33**(10), 476–479 (1980) [Ger., *Chem. Abstr.*, **94**(15), 121266r].

S. Bolanca and A. Golubovic, "The determination of ink composition by means of the simplex method," *Hem. Ind.*, **34**, 168–172 (1980) [Serbo-Croatian, *Chem. Abstr.*, **93**, 96911s].

J. Borszeki, K. Doerffel, and E. Gegus, "Application of experimental planning methods in chemical research. III. Optimization of calcium atomic absorption determination using the simplex method," *Magy. Kem. Foly.*, **86** (1980) 207–210 [Hung., *Chem. Abstr.*, **93**(4), 36232m].

D. B. Chestnut and F. W. Whitehurst, "A simplex optimized INDO calculation of ^{13}C chemical shifts in hydrocarbons," *J. Comput. Chem.*, **1**, 36–45 (1980).

F. L. Chubb, J. T. Edward, and S. C. Wong, "Simplex optimization of yields in the Bucherer–Bergs reaction," *J. Org. Chem.*, **45**, 2315–2320 (1980).

L. Ebdon, M. R. Cave, and D. J. Mowthorpe, "Simplex optimization of inductively coupled plasmas," *Anal. Chim. Acta*, **115**, 179–187 (1980).

R. Fletcher, *Practical Methods of Optimization. Volume I: Unconstrained Optimization*, John Wiley, New York, 1980, pp. 14–16.

S. Fujii and S. Nakai, "Optimization of data transformations for linearization," *Can. Inst. Food Sci. Technol. J.*, **13**(4), 188–191 (1980).

C. Hendrix, "Through the response surface with test tube and pipe wrench," *CHEMTECH*, **10**(8), 488–497 (1980).

F. Hsu, J. Anderson, and A. Zlatkis, "A practical approach to optimization of a selective gas-chromatographic detector by a sequential simplex method," *J. High Res. Chrom. Chrom. Commun.*, **3**, 648–650 (1980).

K. Hyakuna and G. L. Samuelson, "Shimplex – automatic control system for Y and C shims," *JEOL News*, **16A**(1), 17–19 (1980).

S. V. Lyubimova, S. V. Mamikonyan, Y. N. Svetailo, and K. I. Shchekin, "Optimization of the composition of substances for an experiment in calibration of a multielement radiometric x-ray analyzer," *Vopr. At. Nauki Tekh., [Ser.]: Radiats. Tekh.*, **19**, 176–178 (1980) [Russ., *Chem. Abstr.*, **95**, 125405d].

A. Mangia, "A new approach to the problem of kaolinite interference in the determination of chrysolite asbestos by means of x-ray diffraction," *Anal. Chim. Acta*, **117**, 337–342 (1980).

R. J. McDevitt and B. J. Barker, "Simplex optimization of the synergic extraction of a bis(diketo)copper(II) complex," *Anal. Chim. Acta*, **122**, 223–226 (1980).

T. Michalowski, A. Rokosz, and E. Wojcik, "Optimization of the conventional method for determination of zinc as 8-hydroxyquinolate in alkaline tartrate medium," *Chem. Anal. (Warsaw)*, **25**, 563–566 (1980) [*Chem. Abstr.*, **94**(16), 131609u].

G. Minkova and V. Baeva, "Optimization of the isomerization of p-(isoamyloxy) anilinium thiocyanate to [p-(isoamyloxy)phenyl] thiourea. I. Experimental optimization by the simplex method," *Khim. Ind. (Sofia)*, **9**, 402–404 (1980) [Bulg., *Chem. Abstr.*, **95**, 24485z].

J. Pozivil, F. Cermak, and J. Michalek, "Optimization of a multistage adiabatic reactor with an exothermic reversible reaction. I. Optimal catalyst distribution in a state in relation to the flow rate and composition of sulfur dioxide," *Sb. Vys. Sk. Chem.-Technol. Praze*, **3**, 65–86 (1980) [Czech., *Chem. Abstr.*, **94**(8), 49629s].

P. B. Ryan, R. L. Barr, and H. D. Todd, "Simplex techniques for nonlinear optimization," *Anal. Chem.*, **52**, 1460–1467 (1980).

E. Shek, M. Ghani, and R. E. Jones, "Simplex search in optimization of capsule formulation," *J. Pharm. Sci.*, **69**, 1135–1142 (1980).

S. K. Silber, R. A. Deans, and R. A. Geanangel, "A comparison of the simplex and Gauss iterative algorithms for curve fitting in Moessbauer spectra," *Comput. Chem.*, **4**, 123–130 (1980).

S. Stieg and T. A. Nieman, "Application of a microcomputer controlled chemiluminescence research instrument to the simultaneous determination of cobalt(II) and silver(I) by gallic acid chemiluminescence," *Anal. Chem.*, **52**, 800–804 (1980).

V. Svoboda, "Search for optimal eluent composition for isocratic liquid column chromatography," *J. Chromatogr.*, **201**, 241–252 (1980).

P. F. A. Van der Wiel, "Improvement of the super-modified simplex procedure," *Anal. Chim. Acta*, **122**, 421–433 (1980).

J. T. Wrobleski and D. B. Brown, "A study of the variable-temperature magnetic susceptibility of two Ti(III) oxalate complexes," *Inorg. Chim. Acta*, **38**, 227–230 (1980).

C. Zimmermann and W. E. Hoehne, "Simplex optimization of a fluorometric determination of the pyruvate kinase and phosphofructoskinase activities from rabbit muscle using fluorescent adenine nucleotides," *Z. Med. Laboratoriumsdiagn.*, **21**, 259–267 (1980) [Ger., *Chem. Abstr.*, **94**(1), 1422s].

1981:

R. L. Birke, M.-Y. Kim, and M. Strassfeld, "Diagnosis of reversible, quasi-reversible, and irreversible electrode processes with differential pulse polarography," *Anal. Chem.*, **53**, 852–856 (1981).

S. Boy and K. Boehme, "Determination of kinetic parameters from TG curves by nonlinear optimization," *Thermochim. Acta*, **46**(3), 229–237 (1981).

M. R. Cave, D. M. Kaminaris, L. Ebdon, and D. J. Mowthorpe, "Fundamental studies of the application of an inductively coupled plasma to metallurgical analysis," *Anal. Proc. (London)*, **18**(1), 12–14 (1981).

M. F. Delaney and F. V. Warren, Jr., "Teaching chemometrics: A course on application of mathematical techniques to chemistry," *J. Chem. Educ.*, **58**(8), 646–651 (1981).

S. N. Deming, "The role of optimization strategies in the development of analytical chemical methods," *Am. Lab.*, **13**(6), 42, 44 (1981).

D. K. Dodoo, and M. Vrchlabsky, "Optimization of the silver diethyldithiocarbamate spectrophotometric method for the determination of arsenic in tap water," *Chem. Anal. (Warsaw)*, **26**(5), 867–876 (1981).

L. Ebdon, "The optimization of an inductively coupled plasma for metallurgical analysis," in R. M. Barnes, Ed., *Proceedings of the International Winter Conference on Developments in Atomic Plasma Spectrochemical Analysis 1980*, Heyden, London, 1981, pp. 94–110 [*Chem. Abstr.*, **96**, 173462].

C. E. Fiori, C. R. Swyt, and K. E. Gorlen, "Application of the top-hat digital filter to a nonlinear spectral unraveling procedure in energy-dispersive x-ray microanalysis," *Microbeam Anal.*, **16**, 320–324 (1981) [*Chem. Abstr.*, **96**(6), 45385c].

C. E. Fiori, R. L. Myklebust, and K. Gorlen, "Sequential simplex: A procedure for resolving spectral interference in energy dispersive x-ray spectrometry," *NBS Spec. Publ. (U.S.)*, **604**, 233–272 (1981) [*Chem. Abstr.*, **95**(8), 70461m].

O. T. Inal, M. Valayapetre, L. E. Murr, and A. E. Torma, "Microstructural and mechanical property evaluation of black-chrome coated solar collectors – II," *Sol. Energy Mater.*, **4**(3), 333–358 (1981).

K. Jung, D. Scholz, and G. Schreiber, "Improved determination of D-glucaric acid in urine," *Clin. Chem. (Winston-Salem, NC)*, **27**(3), 422–426 (1981).

M. Matsumura, T. Imanaka, T. Yoshida, and H. Taguchi, "Optimal conditions for production of cephalosporin C in fed-batch culture," in M. Moo-Young, C. W. Robinson, and C. Vezina, Eds., *Advances in Biotechnology: Proceedings of the 6th Annual International Fermentation Symposium*, Volume 1, Pergamon, Toronto, Ontario, 1981, pp. 297–302, [*Chem. Abstr.*, **96**(19), 160714t].

M. Matsumura, T. Imanaka, T. Yoshida, and H. Taguchi, "Modelling of cephalosporin C production and its application to fed-batch culture," *J. Ferment. Technol.*, **59**(2), 115–123 (1981).

R. J. Matthews, S. R. Goode, and S. L. Morgan, "Characterization of an enzymatic determination of arsenic using response surface methodology," *Anal. Chim. Acta*, **133**, 169–182 (1981).

M. Otto and G. Werner, "Optimization of a kinetic-catalytic method by the use of a numerical model and the simplex method," *Anal. Chim. Acta*, **128**, 177–183 (1981).

O. V. Sakhartova, V. Sates, J. Freimanis, and A. Avots, "Chromatography of prostaglandins, their analogs and precursors. I. Gas chromatographic control of the stages of 2-(6-carboethoxyhexyl)-6-endo-vinylbicyclo[3.1=0]hexan-1-one synthesis," *Latv. PSR Zinat. Akad. Vestis, Kim Ser.*, **4**, 414–420 (1981) [Russ., *Chem. Abstr.*, **95**, 180312w].

H. P. Schwefel, *Numerical Optimization of Computer Models*, John Wiley, New York, 1981, pp. 57–65.

M. M. Siegel, "The use of the modified simplex method for automatic phase correction in Fourier-transform nuclear magnetic resonance spectroscopy," *Anal. Chim. Acta*, **133**, 103–108 (1981).

G. L. Silver, "Space modification: An alternative approach to chemistry problems involving geometry," *J. Comput. Chem.*, **2**(4), 478–482 (1981).

O. Strouf and J. Fusek, "Simplex search for mathematical representation of chemical class structure," *Coll. Czech. Chem. Commun.*, **46**, 58–64 (1981).

M. Suchanek, E. Svecova, and L. Sucha, "Optimization of methods for the determination of chlorides in the presence of excess boric acid," *Sb. Vys. Sk. Chem.-Technol. Praze, Anal. Chem.*, **H-16**, 65–75 (1981) [*Chem. Abstr.*, **97**(6), 48854m].

I. Taufer and J. Tauferova, "Experiment planning in determining the operational regime of an atomic absorption spectrophotometer by the simplex method," *Chem. Prum.*, **31** 16–20 (1981) [Czech., *Chem. Abstr.*, **94**, 95145d].

S. P. Terblanche, K. Visser, and P. B. Zeeman, "The modified sequential simplex method of optimization as applied to an inductively coupled plasma source," *Spectrochim. Acta, Part B*, **36B**, 293–297 (1981).

A. S. Ward, "Optimization of processes and products in filtration and separation," *Filtech Conference Proceedings*, Uplands Press Ltd., Croydon, UK, 1981, pp. 365–371.

R. L. White, G. N. Giss, G. M. Brissey, and C. L. Wilkins, "Comparison of methods for reconstruction of gas chromatograms from interferometric gas chromatography/infrared spectrometry data," *Anal. Chem.*, **53**, 1778–1782 (1981).

K. Yamaoka, Y. Tanigawara, T. Nakagawa, and T. Uno, "A pharmacokinetic analysis program (MULTI) for microcomputer," *J. Pharmacobio-Dyn.*, **4**(11), 879–885 (1981) [*Chem. Abstr.*, **96**(7), 45844b].

T. Yoshida, M. Sueki, H. Taguchi, S. Kulprecha, and N. Nilubol, "Modelling and optimization of steroid transformation in a mixed culture," *Eur. J. Appl. Microbiol. Biotechnol.*, **11**, 81–88 (1981).

T. Yoshida, H. Taguchi, S. Kulprecha, and N. Nilubol, "Kinetics and optimization of steroid transformation in a mixed culture," in C. Vezina and K. Singh, Eds., *Advances in Biotechnology: Proceedings of the 6th Annual International Fermentation Symposium*, Volume 3, Pergamon, Toronto, 1981, pp. 501–506 [*Chem. Abstr.*, **96**(17), 141090y].

S. P. Terblanche, K. Visser, and P. B. Zeeman, "The modified sequential simplex method of optimization as applied to an inductively coupled plasma source," *Spectrochim. Acta, Part B*, **36B**(4), 293–297 (1981).

V. B. Zierenberg and H. Stricker, "Comparison of different optimizing methods on galenic developmental problems. Part I: Theoretical examples," *Pharm. Ind.*, **43**(8), 777–781 (1981).

1982:

E. R. Aberg and A. G. T. Gustavsson, "Design and evaluation of modified simplex methods," *Anal. Chim. Acta*, **144**, 39–53 (1982).

J. C. Berridge, "Optimization of high-performance liquid chromatographic separations with the aid of a microcomputer," *Anal. Proc. (London)*, **19**(10), 472–475 (1982).

J. C. Berridge, "Unattended optimisation of reversed-phase high-performance liquid chromatographic separations using the modified simplex algorithm," *J. Chromatogr.*, **244**, 1–14 (1982).

J. C. Berridge, "Unattended optimization of normal phase high-performance liquid chromatography separations with a microcomputer controlled chromatograph," *Chromatographia*, **16**, 172–174 (1982).

D. Betteridge, A. P. Wade, E. A. Neves, and I. Gutz, "Computer assisted optimization of sensitivity of the catalytic polarographic wave of uranyl/nitrate system," *An. Simp. Bras. Eletroquim. Eletroanal, 3rd*, Volume 2, 411–419 (Dep. Quim. Univ. Fed. São Carlos, São Carlos, Brazil) (1982).

C. Bindschaedler and R. Gurny, "Optimization of pharmaceutical formulations by the simplex method with a TI 59 calculator," *Pharm. Acta Helv.*, **57**(9), 251–255 (1982) [*Chem. Abstr.*, **97**(20), 168837e].

J. J. Brocas, "Simplex technique for nonlinear optimization: its application to oil analysis by inductively coupled plasma emission spectrometry," *Analusis*, **1982**, 387–389 (1982) (Fr.).

H. N. Cheng, "Markovian statistics and simplex algorithm for carbon-13 nuclear magnetic resonance spectra of ethylene-propylene copolymers," *Anal. Chem.*, **54**, 1828–1833 (1982).

L. Ebdon, R. W. Ward, and D. A. Leathard, "Development and optimization of atom cells for sensitive coupled gas chromatography-flame atomic absorption spectrometry," *Analyst (London)*, **107**, 129–143 (1982).

D. M. Fast, P. H. Culbreth, and E. J. Sampson, "Multivariate and univariate optimization studies of liquid-chromatographic separation of steroid mixtures," *Clin. Chem. (Winston-Salem, NC)*, **28**(3), 444–448 (1982).

M. Forina and C. Armanino, "Eigenvector projection and simplified nonlinear mapping of fatty acid content of Italian olive oils," *Ann. Chim. (Rome)*, **72**, 127–141 (1982).

L. de Galan and G. R. Kornblum, "Computer control of monochromators in atomic spectrometry," *Chem. Mag. (Rijswijk, Neth.)*, 240–241 (1982) [*Chem. Abstr.*, **97**(22), 192324f].

M. Gassiot, X. Tomas, F. Broto, L. G. Sabate, and G. Codinas, "Algorithms for the quantitative determination of PCB's in chromatograms of organochlorinated contaminants: Application to samples of sediments and marine organisms," *Pergamon Ser. Environ. Sci.*, **7** (Anal. Tech. Environ. Chem. 2), 249–258 (1982).

H. P. Huang and Y. C. Chao, "Optimal tuning of a practical digital PID controller," *Chem. Eng. Commun.*, **18**(1–4), 51–61 (1982).

G. F. Kirkbright and S. J. Walton, "Optical emission spectrometry with an inductively coupled radiofrequency argon plasma source and direct sample introduction from a graphite rod," *Analyst (London)*, **107**, 276–281 (1982).

J. J. Leary, A. E. Brooks, A. F. Dorrzapf, Jr., and D. W. Golightly, "An objective function for optimization techniques in simultaneous multiple-element analysis by inductively coupled plasma spectrometry," *Appl. Spectrosc.*, **36**, 37–40 (1982).

G. C. Levy, D. J. Craik, B. Norden, M. T. Pahn Viet, and A. Dekmezian, "The 6-X-benzonorbornyl system: A new motional dynamics probe," *J. Am. Chem. Soc.*, **104**(1), 25–28 (1982).

S. Nakai, "Comparison of optimization techniques for application to food product and process development," *J. Food Sci.*, **47**(1), 144–152, 157 (1982).

M. Roura, M. Baucells, G. Lacort, and G. Rauret, "Determination of aluminum in river water by the inductively coupled plasma-atomic spectroscopic (ICP-AES) technique," *Pergamon Ser. Environ. Sci.*, **7** (Anal. Tech. Environ. Chem. 2), 377–380 (1982).

S. Suzuki and K. Fukunishi, "An autotuning method for control system parameters in nuclear power plants," *Nucl. Technol.*, **58**(3), 379–387 (1982).

B. G. M. Vandeginste, "Optimization of analytical information," *Trends Anal. Chem.*, **1**(9), 210–215 (1982).

A. S. Ward, "Application of optimization methods of filtration and separation," *Filtr. Sep.*, **19**(4), 326, 328, 352 (1982).

G. Wuensch, N. Czech, and G. Hegenberg, "Determination of tungsten with the capacitively coupled microwave plasma (CMP). Optimization of a CMP using factorial design and simplex method," *Z. Anal. Chem.*, **310**, 62–69 (1982) [Ger., *Chem. Abstr.*, **96**, 192515y].

W. Wegscheider, E. P. Lankmayr, and K. W. Budna, "A chromatographic response function for automated optimization of separations," *Chromatographia*, **15**, 498–504 (1982).

V. B. Zierenberg, A. R. Gupte, H. Harwalik, and H. Stricker, "Comparison of different optimization methods for pharmaceutical development problems. Part 2: Practical example from tabletting technology," *Pharm. Ind.*, **44**(7), 741–744 (1982).

1983:

R. M. Barnes and P. Fodor, "Analysis of urine using inductively-coupled plasma emission spectroscopy with graphite rod electrothermal vaporization," *Spectrochim. Acta*, **38B**, 1191–1202 (1983).

Y. I. Benshtein, "Determination of the optimal composition of clinker using a sequential simplex method," *Tsement*, (7), 16–17 (1983) [Russ., *Chem. Abstr.*, **99**(14), 109757x].

J. C. Berridge, "High-performance liquid chromatography method development by microcomputer using linear and simplex optimization," *Anal. Proc. (London)*, **20**(1), 29–32 (1983).

D. Betteridge, T. J. Sly, A. P. Wade, and J. E. W. Tillman, "Computer-assisted optimization for flow injection analysis of isoprenaline," *Anal. Chem.*, **55**, 1292–1299 (1983).

G. Brink, L. Glasser, R. A. Hasty, and P. Wade, "Numerical optimization on a microcomputer," *J. Chem. Educ.*, **60**(7), 564 (1983).

R. Carlson, T. Lundstedt, R. Phan Tan Luu, and D. Mathieu, "On the necessity of using multivariate methods for optimization in synthetic chemistry. An instructive example with the Willgerodt reaction," *Nouv. J. Chim.*, **7**(5), 315–319 (1983) [*Chem. Abstr.*, **99**(15), 121436z].

R. Carlson, A. Nilsson, and M. Stroemqvist, "Optimum conditions for enamine synthesis by an improved titanium tetrachloride procedure," *Acta Chem. Scand., Ser. B.*, **B37**, 7–13 (1983).

R. Cela and J. A. Perez-Bustamante, "Resolution of overlapping peaks in HPLC by means of the simplex algorithm," *Comp. Applic. Lab. (CAL)*, **2**, 137–144 (1983) [*Chem. Abstr.*, **99**(24), 205304t].

S. N. Deming and S. L. Morgan, "Teaching the fundamentals of experimental design," *Anal. Chim. Acta*, **150**, 183–198 (1983).

D. L. Dunn and R. E. Thompson, "Reversed-phase high-performance liquid chromatographic separation for pilocarpine and isopilocarpine using radial compression columns," *J. Chromatogr.*, **264**, 264–271 (1983).

D. M. Fast, E. J. Sampson, V. S. Whitner, and M. Ali, "Creatine kinase response surfaces explored by use of factorial experiments and simplex maximization," *Clin. Chem. (Winston-Salem, NC)*, **29**, 793–799 (1983).

M. Forina, C. Armanino, S. Lanteri, and C. Calcagno, "Simplified nonlinear mapping of analytical data," *Ann. Chim. (Rome)*, **73**(11–12), 641–657 (1983) [*Chem. Abstr.*, **100**(6), 44510d].

M. Forina, C. Armanino, S. Lanteri, and E. Tiscornia, "Classification of olive oils from their fatty acid patterns," in H. Martens and H. Russwurm, Jr., Eds., *Food Research and Data Analysis*, Applied Science Publ., Barking, U.K., 1983, pp. 189–214.

J. Gonzalez, F. Berti, and J. L. Lamazares, "Simplex optimization of experimental factors in atomic absorption spectroscopy. Part 1," *Rev. Cubana Fis.*, **3**(2), 141–153 (1983) [*Chem. Abstr.*, **101**(12), 103052b].

J. Gonzalez, F. Berti, and M. Hernandez Martinez, "Simplex optimization of experimental factors in atomic spectroscopy. Part 2. Applications in the atomic emission spectroscopy," *Rev. Cubana Fis.*, **3**(3), 111–123 (1983).

P. L. Gould and M. Goodman, "Simplex optimization of the solubility of caffeine in parenteral co-solvent systems," *J. Pharm. Pharmacol.*, **35**, Suppl., 3P (1983).

A. Gruberova and M. Suchanek, "Optimization procedures in extraction-spectrophotometric determinations," *Sb. Vys. Sk. Chem.-Technol. Praze, Anal. Chem.*, **H-18**, 167–177 (1983) [*Chem. Abstr.*, **101**(24), 221456j].

S. L. Harper, J. F. Walling, D. M. Holland, and L. J. Pranger, "Simplex optimization of multielement ultrasonic extraction of atmospheric particulates," *Anal. Chem.*, **55**, 1553–1557 (1983).

S. A. Howard and R. L. Snyder, "An evaluation of some profile models and the optimization procedures used in profile fitting," *Adv. X-Ray Anal.*, **26**, 73–80 (1983) [*Chem. Abstr.*, **99**(14), 115038e].

T. A. H. M. Janse, P. F. A. Van der Wiel, and G. Kateman, "Experimental optimization procedures in the determination of phosphate by flow-injection analysis," *Anal. Chim. Acta*, **155**, 89–102 (1983)].

M. H. Kim and R. L. Birke, "Differential pulse polarography for a first-order catalytic process," *Anal. Chem.*, **55**, 522–527 (1983).

M. H. Kim and R. L. Birke, Differential pulse polarography for an electrode process with a prior chemical equilibrium," *Anal. Chem.*, **55**, 1735–1741 (1983).

K. Kimoto, "Water absorption and Donnan equilibriums of perfluoro ionomer membranes for the chlor-alkali process," *J. Electrochem. Soc.*, **130**(2), 334–341 (1983).

J. Klavins, I. O. Dreier, and A. P. Raman, "Use of mathematical methods for optimization of compositions of low-melting glazes," *Neorg. Stekla, Pokrytiya Mater.*, **6**, 102–108 (1983) [Russ., *Chem. Abstr.*, **101**(14), 115607z].

E. J. Knudson, and K. J. Siebert, "High performance liquid chromatographic analysis of hop bittering components in hops, hop extracts, wort, and beer," *J. Am. Soc. Brew. Chem.*, **41**(2), 51–57 (1983).

G. R. Kornblum, J. Smeyers-Verbeke, Y. Michotte, A. Klok, D. L. Massart, and L. de Galan, "Optimization and long term stability of the induction coupled plasma in a synthetical biological matrix," in P. Braetter and P. Schramel, Eds., *Trace Element–Analytical Chemistry in Medicine and Biology*, Volume 2, Walter de Gruyter, Berlin, 1983, pp. 1161–1173.

G. Kostal, G. B. Ashe, and M. P. Eastman, "Generation of component set representations for rocks using the simplex approach," *J. Chem. Geol.*, **39**(4), 347–356 (1983).

D. G. Leggett, "Instrumental simplex optimization. Experimental illustrations for an undergraduate laboratory course," *J. Chem. Educ.*, **60**(9), 707–710 (1983).

H. Lin, D. Qiu, and X. Chang, "Simplex optimization of acetylacetone method for spectrophotometric determination of formaldehyde," *Hunan Daxue Xuebao*, **10**(2), 106–110 (1983) (Ch.) [*Chem. Abstr.*, **100**(24), 202846z].

H. Masson, "A Monte-Carlo approach to the interpretation of bubble probe signals in freely bubbling fluid beds," *Chem. Eng. Commun.*, **21**, 311–328 (1983).

J. T. McCaffrey and R. G. Michel, "Carbon furnace for sample introduction into a metastable nitrogen plasma," *Anal. Chem.*, **55**(13), 2175–2179 (1983).

J. H. Nickel and S. N. Deming, "Use of the sequential simplex algorithm for improved separations in automated liquid chromatographic methods development," *LC Mag.*, **1**(7), 414–417 (1983).

N. V. Nikitin and B. V. Sazykin, "Optimization of heterogeneous shielding compositions based on a sequential simplex method," *At. Energ.*, **55**(1), 48–49 (1983) [Russ., *Chem. Abstr.*, **99**(16), 129927f].

J. J. O'Dea, J. Osteryoung, and R. A. Osteryoung, "Square wave voltammetry and other pulse techniques for the determination of kinetic parameters. The reduction of zinc(II) at mercury electrodes," *J. Phys. Chem.*, **87**(20), 3911–3918 (1983).

M. Otto, G. Schoebel, and G. Werner, "Kinetic-catalytic determination of vanadium in nonaqueous solvents," *Anal. Chim. Acta*, **147**, 287–292 (1983).

M. Otto, J. Rentsch, and G. Werner, "Optimized spectrophotometric determination of trace cobalt and manganese by their catalysis of Tiron-hydrogen peroxide reaction," *Anal. Chim. Acta*, **147**, 267–275 (1983).

N. V. Pershin and V. I. Mosichev, "Planning of optimal systems of standards for x-ray spectrometric analysis of multicomponent materials," *Zavod. Lab*, **49**(12), 34–39 (1983) [Russ., *Chem. Abstr.*, **100**(12), 95633d].

L. G. Sabate, A. M. Diaz, X. M. Tomas, and M. M. Gassiot, "Computer-assisted optimization in HPLC," *J. Chromatogr. Sci.*, **21**(10), 439–443 (1983).

M. Sisido, S. Egusa, and Y. Imanishi, "One-dimensional aromatic crystals in solution. 1. Synthesis, conformation, and spectroscopic properties of poly(L-1-pyrenylalanine)," *J. Am. Chem. Soc.*, **105**, 1041–1049 (1983).

M. Sisido, S. Egusa, and Y. Imanishi, "One-dimensional aromatic crystals in solution. 2. Synthesis, conformation, and spectroscopic properties of poly(L-2-naphthylalanine)," *J. Am. Chem. Soc.*, **105**, 4077–4082 (1983).

P. R. Sthapit, J. M. Ottaway, and G. S. Fell, "Determination of lead in blood by flame atomic-fluoresence spectrometry," *Analyst (London)*, **108**, 235–243 (1983).

T. Sunden, M. Lindgren, A. Cedergren, and D. D. Siemer, "Separation of sulfite, sulfate, and thiosulfate by ion chromatography with gradient elution," *Anal. Chem.*, **55**, 2–4 (1983).

P. F. A. Van der Wiel, R. Maassen, and G. Kateman, "The symmetry-controlled simplex optimization procedure," *Anal. Chim. Acta*, **153**, 83–92 (1983).

A. P. Wade, "Optimization of flow injection analysis and polarography by the modified simplex method," *Anal. Proc. (London)*, **20**(10), 523–527 (1983).

F. H. Walters and K. B. Griffin, "Studies on the reaction of o-phthaldialdehyde with triphenylmethanethiol and amino acids," *Anal. Lett.*, **16**(A6), 485–490 (1983).

D. E. Weisshaar, and D. E. Tallman, "Chronoamperometric response at carbon-based composite electrodes," *Anal. Chem.*, **55**, 1146–1151 (1983).

M. Xu and Y. He, "Simplex optimization of W-PR-CTMAB spectrophotometric determination of tungsten," *Zhongnan Kuangye Xueyuan Xuebao*, (2), 77–82 (1983) [Ch., *Chem. Abstr.*, **99**(22), 186604x].

1984:

A. Adjemian and P. Colombe, "Optimization of a Bayer flowsheet by successive simulations automatically generated," *Light Met. (Warrendale, PA)*, 3–11 (1984) [*Chem. Abstr.*, **100**(24), 194399p].

A. Aliakbar and M. Popl, "The determination of trace amounts of heavy metals in foodstuffs by anodic stripping voltammetry: Optimization of chemical factors," *Collect. Czech. Chem. Commun.*, **49**(5), 1140–1148 (1984).

L. Backman and V. P. Shanbhag, "Simplex optimization in biochemistry: Application of the method in two-phase partition," *Anal. Biochem.*, **138**(2), 372–379 (1984).

J. C. Berridge, "Automated multiparameter optimization of high-performance liquid chromatographic separations using the sequential simplex procedure," *Analyst (London)*, **109**(3), 291–293 (1984).

J. C. Berridge and E. G. Morrisey, "Automated optimization of reversed-phase high-performance liquid chromatography separations," *J. Chromatogr.*, **316**, 69–79 (1984).

D. Betteridge, A. F. Taylor, and A. P. Wade, "Optimization of conditions for flow injection analysis," *Anal. Proc. (London)*, **21**(10), 373–375 (1984).

D. D. Burgess and P. Hayumbu, "Simplex optimization by advance prediction for single-element instrumental neutron activation analysis," *Anal. Chem.*, **56**, 1440–1443 (1984).

M. S. Caceci and W. P. Cacheris, "Fitting curves to data," *BYTE*, 340–362 (May, 1984).

R. Carlson, A. Nilsson, and T. Lundstedt, "Chemical experiment planning. Part 3. Experimental design–strategies for optimization," *Kem. Tidskr.*, **96**(5), 48–50, 53, 55–56 (1984) (Swed.) [*Chem. Abstr.*, **101**(5), 37747d].

R. Cela and J. A. Perez-Bustamante, "El metodo simplex y sus aplicaciones en quimica analitica," *Quim. Anal. (Barcelona)*, **3**(2), 87–128 (1984).

L. Chen, L. Zhang, and S. Sun, "Determination of conditions for determining platinum, palladium, and gold by atomic absorption spectrometry using the simplex optimization method," *Fenxi Huaxue*, **12**(2), 124–127 (1984) (Ch.) [*Chem. Abstr.*, **100**(20), 167324d].

Z. Chen and Q. Ru, "An empirical flotation model and optimal selection of operating variables," *Youse Jinshu*, **36**(2), 34–41 (1984) [Ch., *Chem. Abstr.*, **101**(26), 233657p].

S. N. Deming, J. G. Bower, and K. D. Bower, "Multifactor optimization of HPLC conditions," in J. C. Gidding, Ed., *Advances in Chromatography*, Volume 24, Chapter 2, pp. 35–53, 1984.

S. N. Deming and S. L. Morgan, *Instrumentune-up–A Computer Program for Improving Instrument Response*, Elsevier Scientific Software, Amsterdam, 1984; IBM PC version, ISBN 0-444-42330-3.

R. R. Eley, "Thermosetting coatings: Analytical and predictive capability by chemorheology," *J. Coatings Technol.*, **56**(718), 49–56 (1984).

M. Feinberg and P. Wirth, "General introduction to optimization in analytical chemistry," *Analusis*, **12**(10), 490–495 (1984) [Fr., *Chem. Abstr.*, **102**(4), 38793y].

A. Fernandez, M. D. Luque de Castro, and M. Valcarcel, "Comparison of flow injection analysis configurations for differential kinetic determination of cobalt and nickel," *Anal. Chem.*, **56**(7), 1146–1151 (1984).

M. Forina and S. Lanteri, "Data analysis in food chemistry," in B. R. Kowalski, Ed., *Chemometrics. Mathematics and Statistics in Chemistry*, D. Reidel, 1984, pp. 305–349.

P. J. Golden and S. N. Deming, "Sequential simplex optimization with laboratory microcomputers," *Lab. Microcomput.*, **3**(2), 44–47 (1984).

P. L. Gould, "Optimization methods for the development of dosage forms," *Int. J. Pharm. Technol. Prod. Manuf.*, **5**(1), 19–24 (1984).

P. L. Gould and I. L. Smales, "Application of sequential simplex optimization to formulation development: Formulation of a 300 mg aspirin tablet," *J. Pharm. Pharmacol.*, **36**, Suppl., 4P (1984).

M. F. Grossi de Sa, C. De Sa, E. R. Pereira de Almeida, W. Barbosa da Cruz, S. Astolfi Filho, and E. Silvano, "Optimization of a protein synthesizing lysate system from *Trypanosoma cruzi*," *Mol. Biochem. Parasitol.*, **10**(3), 347–354 (1984).

E. Keshavarz and S. Nakai, "Utilization of resin-neutralized whey in making cake and bread," *Can. Inst. Food Sci. Technol. J.*, **17**(2), 107–110 (1984).

M. E. Koehler, A. F. Kah, C. M. Neag, T. F. Niemann, F. B. Malihi, and T. Provder, "Computer automation of the dynamic mechanical analyzer and its application to cure kinetics studies and dynamic mechanical property analysis of organic coatings," in J. F. Johnson and P. S. Gill, Eds., *Analytical Calorimetry*, Volume 5, Plenum, New York, 1984, pp. 361–375.

M. E. Lacy, R. K. Crouch, and C. F. Lam, "Computer modelling of the recombination reaction of rhodopsin," *Comput. Biol. Med.*, **14**(4), 403–410 (1984).

E. P. Lankmayr and W. Wegscheider, "Optimization strategies for chromatographic analysis of multi-component mixtures," in E. Reid and I. D. Wilson, Eds., *Drug Determinations in Therapeutic and Forensic Contexts*, Plenum, New York, 1984, pp. 81–89; *Methodol. Surv. Biochem. Anal.*, **14**, 81–89 (1984) [*Chem. Abstr.*, **102**(23), 200563x].

D. Liang, Z. Liang, and W. Wang, "Determination of trace cobalt in sediments by flameless atomic absorption spectrophotometry–application of modified simplex optimization method," *Redai Haiyang*, **3**(2), 7–15 (1984) [Ch., *Chem. Abstr.*, **105**(10), 84817u].

D. Liang and W. Wang, "Simplex optimization and its application in analytical chemistry," *Huaxue Tongbao*, (2), 30–34 (1984) [Ch., *Chem. Abstr.*, **101**(16), 142872p].

D. Liang, W. Wang, and Z. Liang, "Computer simplex method of optimization for the determination of lithium in seawater by flame atomic absorption spectroscopy," *Haiyang Yu Huzhao*, **15**(2), 127–136 (1984) [Ch., *Chem. Abstr.*, **101**(16), 136690m].

S. J. Lyle and N. A. Za'tar, "Spectrofluorometric determination of terbium as its ternary complex with EDTA and Tiron. Compositional studies, optimization of fluorescence output and conversion to a flow system," *Anal. Chim. Acta*, **162**, 305–313 (1984).

D. G. McMinn, R. L. Eatherton, and H. H. Hill, Jr., "Multiple-parameter optimization of a hydrogen-atmosphere flame ionization detector," *Anal. Chem.*, **56**(8), 1293–1298 (1984).

A. Montaser, G. R. Huse, R. A. Wax, S. K. Chan, D. W. Golightly, J. S. Kane, and A. F. Dorrzapf, Jr., "Analytical performance of a low-gas-flow torch optimized for inductively coupled plasma atomic emission spectrometry," *Anal. Chem.*, **56**(2), 283–288 (1984).

G. L. Moore, P. J. Humphries-Cuff, and A. E. Watson, "Simplex optimization of a nitrogen-cooled argon inductively coupled plasma for multielement analysis," *Spectrochim. Acta, Part B*, **39B**, 915–929 (1984).

S. Nakai, K. Koide, and K. Eugster, "A new mapping super-simplex optimization for food product and process development," *J. Food Sci.*, **49**(4), 1143–1148, 1170 (1984).

C. Pedregal, G. G. Trigo, M. Espada, D. Mathieu, R. Phan Tan Luu, C. Barcelo, J. Lamarca, and J. Elguero, "Fractional factorial design to study the Bucherer–Bergs reaction: Synthesis of the cyclohexane spirohydantoin," *J. Heterocycl. Chem.*, **21**(5), 1527–1531 (1984) [Fr., *Chem. Abstr.*, **102**(15), 131961m].

C. Porte, W. Debreuille, and A. Delacroix, "The simplex method and its derivations. Application to optimization in the chemical development laboratory," *Actual Chim.*, (8), 45–54 (1984) [Fr., *Chem. Abstr.*, **102**(16), 134234n].

F. Puttemans and D. L. Massart, "Total and differential determination of arsenic in water and biological materials by electrothermal atomic absorption spectroscopy," *Mikrochim. Acta*, **1**(3–4), 261–270 (1984).

J. Rafel and J. Lema, "Simplification of experimental designs by means of experimental planning. LCHR application to resolution optimization," *Afinidad*, **41**(389), 30–34 (1984) [Span., *Chem. Abstr.*, **100**(26), 216141c].

M. Roura, G. Rauret, J. R. Ivern, and P. Sanchez, "Simplex optimization applied to some analytical problems," *Quim. Anal. (Barcelona)*, **3**(4), 299–310 (1984) [Span., *Chem. Abstr.*, **103**(6), 47426d].

I. B. Rubin, "Determination of optimum pH for the analysis of inorganic phosphate mixtures by phosphorus-31 nuclear magnetic resonance spectroscopy by a simplex procedure," *Anal. Lett.*, **17**(A11), 1259–1267 (1984).

L. G. Sabate and X. Tomas, "Optimization of the mobile phase in TLC by the simplex method. Part 1. Mixtures of solvents," *J. High Res. Chrom. Chrom. Commun.*, **7**(2), 104–106 (1984).

P. Schramel and L. Q. Xu, "Further investigation of an argon-hydrogen plasma in ICP-spectroscopy," *Fresenius' Z. Anal. Chem.*, **319**(3), 229–239 (1984).

F. Steiglich, R. Stahlberg, W. Kluge, "Precision of flame-atomic-absorption-spectrometric determination of main components. 4. Optimization of the cobalt, copper, and chromium determination by applying statistical methods and the SIMPLEX method," *Fresenius' Z. Anal. Chem.*, **317**(5), 527–538 (1984) [Ger., *Chem. Abstr.*, **100**(22), 184984m].

Y. Takahashi, Y. Miyashita, H. Abe, and S. Sasaki, "A new approach for ordered multicategorical classification using simplex technique," *Bunseki Kagaku*, **33**(11), E487–E494 (1984).

Y. Takahashi, Y. Miyashita, Y. Tanaka, H. Hayasaka, H. Abe, and S. Sasaki, "Discriminative structural analysis using pattern recognition techniques in the structure-taste problem of perillartines," *J. Pharm. Sci.*, **73**(6), 737–741 (1984).

P. Victory, R. Nomen, M. Garriga, X. Tomas, and L. G. Sabate, "Utilization of the simplex method of optimization in the preparation of 5-cyano-3-methyl-6-methoxy-1,2,3,4-tetrahydropyridin-2-one," *Afinidad*, **41**(391), 241–243 (1984) [Span., *Chem. Abstr.*, **101**(17), 151724u].

C. P. Wang, D. T. Sparks, S. S. Williams, and T. L. Isenhour, "Comparison of methods for reconstructing chromatographic data from liquid chromatography Fourier transform infrared spectrometry," *Anal. Chem.*, **56**(8), 1268–1272 (1984).

F. H. Walters and S. N. Deming, "A two-factor simplex optimization of a programmed temperature gas chromatographic separation," *Anal. Lett.*, **17**(A19), 2197–2203 (1984).

M. Xu, "Simplex optimization in analytical chemistry," *Fenxi Huaxue*, **12**(2), 151–157 (1984) [Ch., *Chem. Abstr.*, **100**(18), 150132s].

1985:

M. Abdullah and H. Haraguchi, "Computer-controlled graphite cup direct insertion device for direct analysis of plant samples by inductively coupled plasma atomic emission spectrometry," *Anal. Chem.*, **57**(11), 2059–2064 (1985).

J. Albort-Ventura, "Aluminum plating by the simplex method," *Pint. Acabados Ind.*, **27**(139), 19–22, 24–26, 28–30, 32–38 (1985) [*Chem. Abstr.*, **103**(22), 185804b].

A. Aliakbar and M. Popl, "Determination of trace quantities of heavy metals in food by anodic voltammetry," *Collect. Czech. Chem. Commun.*, **50**(5), 1141–1146 (1985).

J. C. Berridge, *Techniques for the Automated Optimization of HPLC Separations*, John Wiley, Chichester, 1985.

D. Betteridge, A. P. Wade, and A. G. Howard, "Reflections on the modified simplex-I," *Talanta*, **32**, 709–722 (1985).

D. Betteridge, A. P. Wade, and A. G. Howard, "Reflections on the modified simplex-II," *Talanta*, **32**, 723–734 (1985).

M. K. L. Bicking and N. A. Adinolfe, "Simplex optimization of extractive alkylation procedures for organic acids in aqueous samples," *J. Chromatogr. Sci.*, **23**(8), 348–351 (1985).

R. Bouche, A. Vanclef, and L. Coclers, "Calculation of the interatomic force constants of pyrazole," *Spectrosc. Lett.*, **18**(6), 473–480 (1985).

D. D. Burgess, "Optimization of multielement instrumental neutron activation analysis," *Anal. Chem.*, **57**, 1433–1436 (1985).

R. Carlson, T. Lundstedt, amd C. Albano, "Screening of suitable solvents in organic synthesis. Strategies for solvent selection," *Acta Chem. Scand., Ser. B*, **B39**(2), 79–91 (1985).

M. De Smet, L. Dryon, and D. L. Massart, "Separation and determination of sulfonamides in a pharmaceutical dosage form by HPLC using optimization procedures," *J. Pharm. Belg.*, **40**(2), 100–106 (1985).

M. F. Delaney and D. M. Mauro, "Extension of multicomponent self-modeling curve resolution based on a library of reference spectra," *Anal. Chim. Acta*, **172**, 193–205 (1985).

S. N. Deming, "Optimization," *J. Res. Natl. Bur. Stand. (U. S.)*, **90**(6), 479–483 (1985).

S. N. Deming and S. L. Morgan, *SIMPLEX-V^{tm}: An Interactive Computer Program for Experimental Optimization*, Statistical Programs, Houston, TX, 1985; IBM-PC Version, ISBN 0-932651-10-0.

S. Egusa, M. Sisido, and Y. Imanishi, "One-dimensional aromatic crystals in solution. 4. Ground- and excited-state interactions of poly(L-1-pyrenylalanine) studied by chiroptical spectroscopy including circularly polarized fluorescence and fluorescence-detected circular dichroism," *Macromolecules*, **18**, 882–889 (1985).

K. E. Gilbert and J. J. Gajewski, "Numerical methods: An implementation of DIF-SUB on an IBM-PC," *Comput. Chem.*, **9**(3), 191–194 (1985).

A. Gustavsson and J-E. Sundkvist, "Design and optimization of modified simplex methods," *Anal. Chim. Acta*, **167**, 1–10 (1985).

X. He, Y. Li, and H. Shi, "Optimization design study of spectrophotometric analysis by the weighted-centroid simplex method," *Fenxi Huaxue*, **13**(5), 344–349 (1985) [Ch., *Chem. Abstr.*, **103**(14), 115136m].

J. V. Ibarra and J. J. Lazaro, "Lignite sulfonation optimized by a modified simplex method," *Ind. Eng. Chem. Prod. Res. Dev.*, **24**(4), 604–607 (1985).

T. Kaneko, B. T. Wu, and S. Nakai, "Selective concentration of bovine immunoglobulins and α-lactalbumin from acid whey using FeCl$_3$," *J. Food Sci.*, **50**(6), 1531–1536 (1985).

J. E. Kipp, "Nonisothermal kinetics—comparison of two methods of data treatment," *Int. J. Pharm.*, **26**(3), 339–354 (1985).

W. P. Kleeman and L. C. Bailey, "Simplex optimization of the blue tetrazolium assay procedure for alpha-ketol steroids," *J. Pharm. Sci.*, **74**(6), 655–659 (1985).

G. Kostal, M. P. Eastman, and N. E. Pingitore, "Geological applications of simplex optimization," *Comput. Geosci.*, **11**(2), 235–247 (1985).

K. Kult and J. Vrbsky, "Optimization of the spectrophotometric determination of germanium with phenylfluorone in the presence of Septonex," *Sb. Vys. Sk. Chem.-Technol. Praze, Anal. Chem.*, **H-20**, 107–118 (1985) [*Chem. Abstr.*, **106**(10), 77892g].

F. Lazaro, M. D. Luque de Castro, and M. Valcarcel, "Sequential and differential catalytic-fluorometric determination of manganese and iron by flow injection analysis," *Anal. Chim. Acta*, **169**, 141–148 (1985).

M. H. Lee, "Application of simplex method in chemistry," *Hwahak Kwa Kongop Ui Chinbo*, **25**(1), 28–31 (1985) [Kor., *Chem. Abstr.*, **103**(2), 8346h].

H. Lin and D. Qui, "Study of reaction of gallium-alizarine complexone-cetylpyridinium bromide by simplex optimization," *Hunan Daxue Xuebao*, **12**(4), 96–100 (1985) [Ch., *Chem. Abstr.*, **104**(16), 141357z].

C. H. Lochmuller, K. R. Lung, and K. R. Cousins, "Applications of optimization strategies in the design of intelligent laboratory robotic procedures," *Anal. Lett.*, **18**(A4), 439–448 (1985).

P. S. Marchetti, P. D. Ellis, and R. G. Bryant, "Cadmium-113 shielding tensors in cadmium-substituted metalloproteins," *J. Am. Chem. Soc.*, **107**(26), 8191–8196 (1985).

W. R. Meier, and E. C. Morse, "A nonlinear, multivariable method for fusion reactor blanket optimization," *Fusion Technol.*, **8**(3), 2665–2680 (1985).

S. Nakai and T. Kaneko, "Standardization of mapping simplex optimization," *J. Food Sci.*, **50**, 845–846 (1985).

M. Novic and J. Zupan, "Hierarchical clustering of carbon-13 nuclear magnetic resonance spectra," *Anal. Chim. Acta*, **177**, 23–33 (1985).

L. R. Parker, Jr., M. R. Cave, and R. M. Barnes, "Comparison of simplex algorithms," *Anal. Chim. Acta*, **175**, 231–237 (1985).

L. R. Parker, Jr., N. H. Tioh, and R. M. Barnes, "Optimization approaches to the determination of arsenic and selenium by hydride generation and ICP-AES," *Appl. Spectrosc.*, **39**(1), 45–48 (1985).

L. Rigal and A. Gaset, "Optimization of the conversion of D-fructose to 5-hydroxymethyl-2-furancarboxaldehyde in a water-solvent-ion exchanger triphasic system. Part II. Search for a local optimum of selectivity by the simplex method," *Biomass*, **8**(4), 267–276 (1985).

S. C. Rutan and S. D. Brown, "Simplex optimization of the adaptive Kalman filter," *Anal. Chim. Acta*, **167**, 39–50 (1985).

B. Schroer and H. P. Hougardy, "Calculating the optimum interference layer contrast," *Prakt. Metallogr.*, **22**(12), 587–596 (1985) [Ger., *Chem. Abstr.*, **104**(8), 54297m].

M. Sisido and Y. Imanishi, "One-dimensional aromatic crystals in solution. 5. Empirical energy and theoretical circular dichroism calculations on helical poly(L-1-pyrenylalanine)," *Macromolecules*, **18**, 890–894 (1985).

M. Sisido, A. Okamoto, and Y. Imanishi, "One-dimensional aromatic crystals in solution. VII. Conformational analysis of poly(β-9-anthrylmethyl L-aspartate) in solution," *Polym. J. (Tokyo)*, **17**(12), 1263–1272 (1985) [*Chem. Abstr.*, **105**(17), 153515h].

J. E. Spencer, G. G. Barna, and D. E. Carter, "Rational process optimization and characterization [in plasma etch processes]," *Proc.-Electrochem. Soc.*, **85-1** (Plasma Process.), 413–423 (1985) [*Chem. Abstr.*, **102**(24), 213349g].

B. To, N. B. Helbig, S. Nakai, and C. Y. Ma, "Modification of whey protein concentrate to stimulate whippability and gelation of egg white," *Can. Inst. Food Sci. Technol. J.*, **18**(2), 150–157 (1985).

M. Villanueva, J. G. Alonso, and F. B. Perez, "Selection of the buffer for the emission spectrometric determination of fifteen elements in nickel oxide (NiO)," *Rev. Cubana Fis.*, **5**(2–3), 11–21 (1985) [Span., *Chem. Abstr.*, **105**(26), 237579v].

F. H. Walters and S. N. Deming, "Window diagrams versus the sequential simplex method: Which is correct?," *Anal. Chim. Acta*, **167**, 361–363 (1985).

L. Xu and P. Schramel, "Modified simplex optimization of argon-hydrogen plasma emission spectroscopy," *Fenxi Huaxue*, **13**(7), 498–503 (1985) [Ch., *Chem. Abstr.*, **103**(26), 226454h].

X. Yin, G. Zhao, B. Yang, and H. Zhang, "Optimum selection for extraction and separation of rare earth elements in soil by PMBP-benzene using the simplex optimization method," *Zhongguo Kexue Jishu Daxue Xuebao*, **15**(2), 164–169 (1985) [Ch., *Chem. Abstr.*, **103**(16), 134117x].

B. Zierenberg, "Use of Nelder-Mead method for the optimization of release parameters in drug polymer systems," *Acta Pharm. Technol.*, **30**(1), 17–21 (1985) [Ger., *Chem. Abstr.*, **103**(4), 27147k].

1986:

D. An and B. Xiang, "Optimization in chromatographic analysis of drugs," *Nanjing Yaoxueyuan Xuebao*, **17**(1), 73–80 (1986) [Ch., *Chem. Abstr.*, **105**(12), 102668c].

P. I. Anagnostopoulou and M. A. Koupparis, "Automated flow-injection phenol red method for determination of bromide and bromide salts in drugs," *Anal. Chem.*, **58**(2), 322–326 (1986).

M. L. Balconi and F. Sigon, "Effect of ammonia in the ion-chromatographic determination of trace anions and optimization of analytical conditions," *Anal. Chim. Acta*, **191**, 299–307 (1986).

P. R. Bedard, and W. C. Purdy, "Simplex data reduction for non-homogenous variance and non-linear equations: An HPLC case study," *Anal. Lett.*, **19**(1–2), 1–11 (1986).

R. L. Belchamber, D. Betteridge, A. P. Wade, A. J. Cruickshank, and P. Davison, "Removal of a matrix effect in ICP-AES multi-element analysis by simplex optimization," *Spectrochim. Acta, Part B*, **41B**(5), 503–505 (1986).

D. Betteridge, T. J. Sly, A. P. Wade, and D. G. Porter, "Versatile automatic development system for flow injection analysis," *Anal. Chem.*, **58**(11), 2258–2265 (1986).

H. Broch and D. Vasilescu, "Conformation and electrostatic properties quantum determination of the new radioprotector and anticancer drug I-102," *Int. J. Quant. Chem., Quant. Biol. Symp.*, **13**, 81–94 (1986).

J. Buitrago, R. Cela, and J. A. Perez-Bustamante, "Automated determination of total polyphenolic indices in white wines by flow injection analysis," *Afinidad*, **43**(406), 530–536 (1986) [Span., *Chem. Abstr.*, **106**(11), 82983s].

D. D. Burgess, "Rotation in simplex optimization," *Anal. Chim. Acta*, **181**, 97–106 (1986).

M. Caballero, R. Cela, and J. A. Perez-Bustamante, "Optimization of analytical foam flotation separations by means of the simplex algorithm," *Sep. Sci. Technol.*, **21**(1), 39–55 (1986).

G. M. Carlson and T. Provder, "Kinetic analysis of consecutive reactions using Nelder–Mead simplex optimization," *Computer Applications in the Polymer Laboratory*, American Chemical Society Symposium Series 313, American Chemical Society, Washington, D.C., 1986, pp. 241–255.

R. Carlson, L. Hansson, and T. Lundstedt, "Optimization in organic synthesis. Strategies when the desired reaction is accompanied by parasitic side reactions," *Acta Chem. Scand., Ser. B*, **B40**(6), 444–452 (1986).

R. C. Carpenter and L. Ebdon, "A comparison of inductively coupled plasma torch–sample introduction configurations using simplex optimization," *J. Anal. At. Spectrom.*, **1**(4), 265–268 (1986).

J. Martinez Calatayud and Campins Falco, "Determination of levamisole hydrochloride with tetraiodomercurate(2-) by a turbidimetric method and flow injection analysis," *Talanta*, **33**(8), 685–687 (1986).

M. R. Cave, "An improved simplex algorithm for dealing with boundary conditions," *Anal. Chim. Acta*, **181**, 107–116 (1986).

R. Cela, C. G. Barroso, C. Viseras, and J. A. Perez-Bustamante, "The PREOPT package for pre-optimization of gradient elutions in high-performance liquid chromatography," *Anal. Chim. Acta*, **191**, 283–297 (1986).

S. L. Chen, J. C. Schug, and J. W. Viers, "The determination of the orientation of anthracene molecules in the unit cell by a refractivity method," *Acta Crystallogr., Sect A; Found. Crystallogr.*, **A42**(3), 137–139 (1986).

V. Chohan, "Nonlinear optimization techniques for accelerator performance improvement on-line: Recent trials and experiment for the CERN antiproton accumulator," *Nucl. Instrum. Methods Phys. Res., Sect. A*, **A247**(1), 190–192 (1986).

S. N. Deming, "Chemometrics: An overview," *Clin. Chem. (Winston-Salem, NC)*, **32**(9), 1702–1706 (1986).

S. N. Deming, "Optimization and experimental design in analytical chemical methods development," Chapter 41 in W. R. Laing, Ed., *Analytical Chemistry and Instrumentation: Proceedings of the 28th Conference on Analytical Chemistry in Energy Technology*, Lewis Publishers, Chelsea, MI, 1986, pp. 293–297.

S. Dong and D. An, "Optimization of HPLC separation of nitroglycerin films by simplex method," *Nanjing Yaoxueyuan Xuebao*, **17**(2), 146–148 (1986) [Ch., *Chem. Abstr.*, **105**(12), 102707q].

E. Elizalde, F. Rueda, M. T. Gutierrez, J. Ortega, and P. Salvador, "Optical constants determination in bilayers: Application to cadmium selenide/titanium and cadmium selenide/tin dioxide," *Sol. Energy Mater.*, **13**(6), 407–418 (1986).

C. E. Goewie, "Optimization of mobile phase composition in liquid chromatography–a survey of most commonly used chemometric procedures," *J. Liquid Chromatogr.*, **9**(7), 1431–1461 (1986).

A. P. Halfpenny and P. R. Brown, "Simplex optimization for the simultaneous HPLC assay of the activities of purine-nucleoside phosphorylase and hypoxanthine-guanine phosphoribosyl transferase," *J. Liquid Chromatogr.*, **9**(12), 2585–2599 (1986).

C. Halldin, S. Stone-Elander, L. Farde, E. Ehrin, and K. J. Fasth, "Preparation of carbon-11-labelled SCH 23390 for the *in vivo* study of dopamine D-1 receptors using positron emission tomography," *Appl. Radiat. Isot.*, **37**(10), 1039–1043 (1986).

S. Hayakawa, Y. Matsuura, R. Nakamura, and Y. Sato, "Effect of heat treatment on preparation of colorless globin from bovine hemoglobin using soluble carboxymethyl cellulose," *J. Food Sci.*, **51**(3), 786–790, 796 (1986).

S. Hayakawa and R. Nakamura, "Optimization approaches to thermally induced egg white lysozyme gel," *Agric. Biol. Chem.*, **50**(8), 2039–2046 (1986).

J. Huang, "A nonlinear calculation of multiple standard addition method in potentiometry with ion-selective electrodes," *Fenxi Huaxue*, **14**(8), 579–583 (1986) [Ch., *Chem. Abstr.*, **105**(26), 237459f].

P. C. Jurs, *Computer Software Applications in Chemistry*, John Wiley, New York, 1986, Chapter 9, pp. 125–140.

E. Kozlowski, T. Gorecki, and M. Bownik, "New high-temperature palladium detectors for oxygen determination. Part III. Optimization of selected operating conditions of the detectors," *Fresenius' Z. Anal. Chem.*, **325**(6), 547–552 (1986).

C. H. Lochmuller and K. R. Lung, "Applications of laboratory robotics in spectrophotometric sample preparation and experimental optimization," *Anal. Chim. Acta*, **183**, 257–262 (1986).

E. Lundberg, D. C. Baxter, and W. Frech, "Constant-temperature atomizer-computer controlled echelle spectrometer system for graphite furnace atomic emission spectrometry," *J. Anal. At. Spectrom.*, **1**(2), 105–113 (1986).

D. M. Mauro and M. F. Delaney, "Resolution of infrared spectra of mixtures by self-modelling curve resolution using a library of reference spectra with simplex-assisted searching," *Anal. Chem.*, **58**(13), 2622–2628 (1986).

R. C. McKellar and H. Cholette, "Determination of the extracellular lipases of *Pseudomonas fluorescens* spp. in skim milk with the beta-naphthyl caprylate assay," *J. Dairy Res.*, **53**(2), 301–312 (1986).

P. Norman and L. Ebdon, "Computer-controlled optimization of an inductively coupled plasma," *Anal. Proc. (London)*, **23**(12), 420–422 (1986).

J. O'Dea, J. Osteryoung, and T. Lane, "Determining kinetic parameters from pulse voltammetric data," *J. Phys. Chem.*, **90**(12), 2761–2764 (1986).

L. Cremades Oliver, A. Mulet Pons, and A. Berna Prats, "Use of the simplex method for solution of chemical engineering problems. Part I. Method and program," *Ing. Quim. (Madrid)*, **18**(212), 225–232 (1986) [Span., *Chem. Abstr.*, **106**(10), 69526a].

G. J. Paquette and R. C. McKellar, "Optimization of extracellular lipase activity from *Pseudomonas fluorescens* using a super-simplex optimization program," *J. Food Sci.*, **51**(3), 655–658 (1986).

D. Perosa, F. Magno, G. Bontempelli, and P. Pastore, "Simplex optimization procedure for evaluating equivalence points in sigmoidal and segmented titration curves," *Anal. Chim. Acta*, **191**, 377–384 (1986).

C. Porte, W. Debreuille, and A. Delacroix, "Simplex. Part 2. Derived methods," *Actual. Chim.*, (6), 1–11 (1986) [Fr., *Chem. Abstr.*, **105**(21), 190171p].

G. S. Pyen, S. Long, and R. F. Browner, "System optimization for the automatic simultaneous determination of arsenic, selenium, and antimony, using hydride generation introduction to an inductively coupled plasma," *Appl. Spectrosc.*, **40**(2), 246–251 (1986).

S. Rapsomanikis, O. F. X. Donard, and J. H. Weber, "Speciation of lead and methyllead ions in water by chromatography/atomic absorption spectrometry after ethylation with sodium tetraethylborate," *Anal. Chem.*, **58**(1), 35–38 (1986).

P. J. Schoenmakers, *Optimization of Chromatographic Selectivity*, Elsevier, Amsterdam, 1986.

M. Sisido and Y. Imanishi, "One-dimensional aromatic crystals in solution. 8. Periodic arrangement of naphthyl chromophores along α-helical polypeptides with varying spacings and orientations," *Macromolecules*, **19**, 2187–2193 (1986).

C. A. R. Skow and M. K. L. Bicking, "Direct alkylation of carboxylic acids in aqueous samples," *Chromatographia*, **21**(3), 157–160 (1986).

H. K. Smith, W. L. Switzer, G. W. Martin, S. A. Benezra, W. P. Wilson, and D. W. Dixon, "A comparison of three automated approaches to HPLC optimization," *J. Chromatogr. Sci.*, **24**(2), 70–75 (1986).

H. A. Spaink, T. T. Lub, G. Kateman, and H. C. Smit, "Automation of the optical alignment of a diode-laser spectrometer by means of simplex optimization," *Anal. Chim. Acta*, **184**, 87–97 (1986).

S. Stieg, "A low-noise simplex optimization experiment," *J. Chem. Educ.*, **63**(6), 546–548 (1986).

X. Tomas, and L. G. Sabate, "The use of simplex optimization in evaluating complex chromatograms of mixtures," *Anal. Chim. Acta*, **191**, 439–443 (1986).

F. Vlacil and V. Hamplova, "Optimization of mobile phase composition in liquid chromatography," *Collect. Czech. Chem. Commun.*, **51**(1), 45–53 (1986) [*Chem. Abstr.*, **104**(8), 161085n].

F. H. Walters and G. Gomez, "A two factor simplex optimization of an ion pair liquid chromatographic separation," *Anal. Lett.*, **19**(17–18), 1787–1792 (1986).

C. Wang, J. Ren, and Z. Zhao, "Application of modified simplex optimization to Bent's limited logarithm method," *Fenxi Huaxue*, **14**(5), 336–340 (1986) [Ch., *Chem. Abstr.*, **105**(8), 67507a].

M. Wolff, D. Kersten, and B. Goeber, "Possibilities for the systematic optimization of HPLC separations," *Pharmazie*, **41**(7), 502–506 (1986) [Ger., *Chem. Abstr.*, **105**(26), 232502e].

O. I. Vershinina, I. L. Nadelyaeva, and V. I. Vershinin, "Simplex optimization with a generalized parameter during kinetic determination of submicrogram amounts of molybdenum," *Izv. Vyssh. Uchebn. Zaved., Khim. Khim. Tekhnol.*, **29**(8), 35–37 (1986) [Russ., *Chem. Abstr.*, **106**(4), 27015c].

J. Yang, B. Xiang, and D. An, "Optimization of the experimental conditions for gas chromatographic analysis of chloramphenicol and its decomposition products by simplex method," *Nanjing Yaoxueyuan Xuebao*, **17**(3), 197–200 (1986) [Ch., *Chem. Abstr.*, **106**(2), 9432g].

R. Zhou, P. Zheng, and X. Wu, "Spectrophotometric determination of tantalum with 9-(4'-aldehydrophenyl)-2,3,7-trihydroxyl-6-fluorone in the presence of CTMAB [cetyltrimethylammonium bromide] with optimization by the simplex method," *Xiyou Jinshu*, **5**(3), 208–214 (1986) [Ch., *Chem. Abstr.*, **105**(26), 237602x].

1987:

S. A. Al-Mashikh and S. Nakai, "Reduction of beta-lactoglobulin content of cheese whey by polyphosphate precipitation," *J. Food Sci.*, **52**(5), 1237–1244 (1987).

J. Alonso, J. Bartroli, J. Coello, and M. Del Valle, "Simultaneous optimization of variables in FIA systems by means of the simplex method," *Anal. Lett.*, **20**(8), 1247–1263 (1987).

T. Ashima, D. L. Wilson, and S. Nakai, "Application of simplex algorithm to flavour optimization based on pattern similarity of GC profiles," in M. Martens, G. A. Dalen, and H. Russwurm, Jr., Eds., *Flavour Science and Technology*, John Wiley, New York, 1987, pp. 19–26.

Q. Bao and Y. Wang, "New simplex optimization technique for curve fitting," *Guangpuxue Yu Guangpu Fenxi*, **7**(3), 70–71 (1987) [Ch., *Chem. Abstr.*, **107**(20), 189684t].

A. L. Beckwith and S. J. Brumby, "Numerical Analysis of EPR spectra. 7. The simplex algorithm," *J. Magn. Res.*, **73**, 252–259 (1987).

W. R. Browett and M. J. Stillman, "Computer aided chemistry. III. Spectral envelope deconvolution based on a simplex optimization procedure," *Comput. Chem.*, **11**(4), 241–250 (1987).

K. W. C. Burton and G. Nickless, "Optimization via simplex. Part I. Background, definitions and a simple application," *Chemom. Intell. Lab. Sys.*, **1**, 135–149 (1987).

M. Caballero, R. Lopez, R. Cela, and J. A. Perez-Bustamante, "Preconcentration and determination of trace metals in synthetic sea water by flotation with inert organic collectors," *Anal. Chim. Acta*, **196**, 287–292 (1987).

J. Martinez Calatayud, P. Campins Falco, and A. Sanchez Sampedro, "Turbidimetric determination of chlorhexidine using flow injection analysis," *Analyst (London)*, **112**(1), 87–90 (1987).

S. N. Deming and S. L. Morgan, *Sequential Simplex Optimization*, ACS Audio Course C-97, American Chemical Society, Washington, D.C., 1987.

B. M. J. De Spiegeleer, P. H. M. De Moerloose, and G. A. S. Slegers, "Criterion for evaluation and optimization in thin-layer chromatography," *Anal. Chem.*, **59**(10), 62–64 (1987).

F. Dondi, Y. D. Kahie, G. Lodi, P. Reschiglian, C. Pietrogrande, C. Bighi, and G. P. Cartoni, "Comparison of the sequential simplex method and linear solvent strength theory in HPLC gradient elution optimization of multicomponent flavonoid mixtures," *Chromatographia*, **23**(11), 844–849 (1987).

L. Ebdon and R. C. Carpenter, "Multielement simplex optimization for inductively coupled plasma/atomic emission spectrometry with a plasma torch having a wide-bore injector tube. Part 1. Conditions for optimum detection limit," *Anal. Chim. Acta*, **200**(1), 551–557 (1987).

L. Ebdon, P. Norman, and S. T. Sparkes, "Simplex optimization of a direct current plasma for atomic emission spectrometry," *Spectrochim. Acta, Part B*, **42B**(4), 619–624 (1987).

L. Ebdon and J. R. Wilkinson, "Direct atomic spectrometric analysis by slurry atomization. Part 1. Optimization of whole coal analysis by inductively coupled plasma atomic emission," *J. Anal. At. Spectrom.*, **2**(1), 39–44 (1987).

M. Forina, S. Lanteri, and C. Armanino, "Chemometrics in food chemistry," *Topics Current Chem.*, **141**, 91–143 (1987).

D. C. Harris, *Quantitative Chemical Analysis*, Freeman, New York, 1987, pp. 674–677.

R. Hebisch, "Computer-assisted simplex optimization of the atomic spectrometric determination of chlorine in organic samples," *Z. Chem.*, **27**(12), 449–450 (1987) [Ger. *Chem. Abstr.*, **108**(20), 179332h].

M. S. Hendrick and R. G. Michel, "Optimization of a direct-current plasma emission echelle spectrometer," *Anal. Chim. Acta*, **192**(2), 183–195 (1987).

X. Hou and P. Xu, "Optimization of analytical conditions for flame atomic absorption spectrometry with modified simplex method," *Huaxue Shijie*, **28**(9), 403–406 (1987) [Ch., *Chem. Abstr.*, **108**(4), 30963g].

J. H. Kalivas, "A simplex optimized inductively coupled plasma spectrometer with minimization of interferences," *Appl. Spectrosc.*, **41**(8), 1338–1342 (1987).

Q. Lu, M. Li, and G. Min, "Simplex optimization of chromatographic conditions for separating aromatic alcohol, aldehyde, and acid," *Sepu*, **5**(2), 98–99 (1987) [Ch., *Chem. Abstr.*, **107**(12), 108542v].

R. Matsuda, M. Ishibashi, and M. Uchiyama, "Simplex optimization of reaction condition using laboratory robotic system," *Yakugaku Zasshi*, **107**(9), 683–689 (1987) [Japan., *Chem. Abstr.*, **108**(4), 27023v].

J. Hernandez Mendez, and F. Becerro Dominguez, "Polarographic study of solutions of cadmium(II) in acetylacetone," *Analyst (London)*, **112**, 231–235 (1987).

G. L. Moore and R. G. Bohmer, "Simplex optimization of a 5-kW nitrogen-cooled argon inductively coupled plasma for maximum signal to background ratios and minimum matrix interference," *J. Anal. At. Spectrom.*, **2**(8), 819–821 (1987).

X. Pan and W. Wang, "Design of computer program for simplex optimization of calorimetric titration data," *Youkuangye*, **6**(1), 46–52 (1987) [Ch., *Chem. Abstr.*, **107**(20), 189798h].

E. L. Plummer, A. A. Liu, and K. A. Simmons, "The application of sequential simplex optimization to pesticide design," in R. Greenhalgh, R. Roberts, and R. Terence, Eds., *Pesticide Science Biotechnology: Proceedings of the 6th International Congress of Pesticide Chemistry*, Blackwell, Oxford, 1987, pp. 65–68 [*Chem. Abstr.*, **107**(25), 231279s].

S. B. Smith, Jr., M. A. Sainz, and R. G. Schleicher, "Optiplex: A multidimensional response surface fitting procedure for optimization of instrumental parameters," *Spectrochim. Acta, Part B*, **42B**(1–2), 323–332 (1987).

P. Werner and H. Friege, "Establishment of optimum conditions for ICP atomic emission spectrometry," *Appl. Spectrosc.*, **41**(1), 32–40 (1987).

K. H. Wong and R. A. Osteryoung, "Real-time simplex optimization of square wave voltammetry," *Electrochim. Acta*, **32**(4), 629–631 (1987).

A. G. Wright, A. F. Fell, and J. C. Berridge, "Sequential simplex optimization and multichannel detection in HPLC: Application to method development," *Chromatographia*, **24**, 533–540 (1987).

1988:

J. Alonso, J. Bartroli, M. Del Valle, A. A. S. C. Machado, and L. M. A. Ribeiro, "Use of a linear function of several variables in simplex optimization as a procedure for assessing analytical versatility in FIA," *J. Chemom.*, **3**(Suppl. A), 249–256 (1988).

M. Caballero, R. Cela, and J. A. Perez-Bustamante, "Solvent sublation of some priority pollutants (phenols)," *Anal. Lett.*, **21**(1), 63–76 (1988).

R. B. Costanzo and E. F. Barry, "Simplex optimization of the alternating-current plasma detector for gas chromatography," *Appl. Spectrosc.*, **42**(8), 1387–1393 (1988).

K. E. Creasy and B. R. Shaw, "Simplex optimization of electroreduction of oxygen mediated by methyl viologen supported on zeolite-modified carbon paste electrode," *Electrochim. Acta*, **33**(4), 551–556 (1988).

B. Deng, P. Chen, and X. Cai, "Simplex optimization of HPLC to separate and determine vitamin E isomers," *Gaodeng Xuexiao Huaxue Xuebao*, **9**(4), 333–336 (1988) [Ch., *Chem. Abstr.*, **109**(11), 91341v].

L. Ebdon and R. Carpenter, "Multielement simplex optimization for inductively-coupled plasma/atomic emission spectrometry with a plasma torch fitted with a wide-bore injector tube. Part 2. Establishment of optimum conditions for minimum interference," *Anal. Chim. Acta*, **209**(1–2), 135–145 (1988).

M. Forina, R. Leardi, C. Armanino, and S. Lanteri, *PARVUS. An Extendable Package of Programs for Data Exploration, Classification, and Correlation*, Elsevier Scientific Software, Amsterdam, 1988; IBM PC version, ISBN 0-444-43012-1.

A. E. Gastaminza, N. N. Ferracutti, and N. M. Rodriguez, "Regiospecific reduction optimization of methyl (E)-2-pentenoate by the multiple-move simplex method," *An. Asoc. Quim. Argent.*, **76**(4), 251–254 (1988) [*Chem. Abstr.*, **110**(21), 192215h].

S. Ghodbane and G. Guiochon, "A simplex optimization of the experimental parameters in preparative liquid chromatography," *Chromatographia*, **26**, 53–59 (1988).

S. Greenfield, M. S. Salman, M. Thomsen, and J. F. Tyson, "A comparison of the alternating variable search and simplex methods of optimization for plasma atomic emission spectrometry," *Anal. Proc. (London)*, **25**(3), 81–85 (1988).

C. He, H. Fan, H. Liu, and L. Zhang, "Selection of operating condition for measuring aldicarb sulfone with gas chromatography-FPD by using modified simplex optimization method," *Sepu*, **6**(3), 175–177 (1988) [Ch., *Chem. Abstr.*, **109**(11), 88035t].

J. Hrabar and Z. Lazic, "Simplex optimization and modeling of the heat treatment of ablation composite materials," *Naucno-Teh. Pregl.*, **38**(3), 13–23 (1988) [Serbo-Croatian, *Chem. Abstr.*, **109**(8), 55856h].

G. W. Jang and K. Rajeshwar, "Computer simulation of differential scanning calorimetry: Influence of thermal resistance factors and simplex optimization of peak resolution," *Anal. Chem.*, **60**(10), 1003–1009 (1988).

Y.-Z. Lee, J. S. Sim, S. Al-Mashikhi, and S. Nakai, "Separation of immunoglobulins from bovine blood by polyphosphate precipitation and chromatography," *J. Agric. Food Chem.*, **36**(5), 922–928 (1988).

L. Ma, S. Huang, and Y. Zhang, "Application of simplex optimization to multicomponent simultaneous determination by spectrophotometry with 2-(5-bromo-2-pyridylazo)-5-(diethylamino)phenol (5-Br-PADAP)," *Yankuang Ceshi*, **7**(3), 166–171 (1988) [Ch., *Chem. Abstr.*, **111**(20), 186464t].

R. Matsuda, M. Ishibashi, and Y. Takeda, "Simplex optimization of reaction conditions with an automated system," *Chem. Pharm. Bull.*, **36**(9), 3512–3518 (1988).

J. Hernandez Mendez, B. Moreno Cordero, J. L. Perez Pavon, and J. Cerda Miralles, "Determination of neodymium with 1-(2-pyridylazo)-2-naphthol by high-order derivative spectrophotometry. Application to the determination of neodymium in glasses," *Analyst (London)*, **113**(3), 429–431 (1988).

G. L. Moore, "Use of simplex optimization to improve the performance of an inductively coupled plasma in atomic emission spectrometry," *TrAC, Trends Anal. Chem.*, **7**(1), 32–35 (1988).

G. R. Phillips and E. M. Eyring, "Error estimation using the sequential simplex method in nonlinear least squares data analysis," *Anal. Chem.*, **60**(8), 738–741 (1988).

T. D. Rhines and M. A. Arnold, "Simplex optimization of a fiber-optic ammonia sensor based on multiple indicators," *Anal. Chem.*, **60**(1), 76–81 (1988).

S. T. Sparkes and L. Ebdon, "Direct atomic spectrometric analysis by slurry atomization. Part 6. Simplex optimization of a direct current plasma for kaolin analysis," *J. Anal. At. Spectrom.*, **3**(4), 563–569 (1988).

A. G. Wright, A. F. Fell, and J. C. Berridge, "Computer-aided optimization with pho-
todiode array detection in HPLC," *Anal. Proc. (London)*, **25**(9), 300–303 (1988).

1989:

S. K. Beh, G. J. Moody, and J. D. R. Thomas, "Modified simplex optimization for
operating an enzyme electrode," *Anal. Proc. (London)*, **26**(8), 290–292 (1989).

J. C. Berridge, "Simplex optimization of high-performance liquid chromatographic
separations," *J. Chromatogr.*, **485**, 3–14 (1989).

S. Brumby, "Exchange of comments on the simplex algorithm culminating in quadratic covergence and error estimation," *Anal. Chem.*, **61**, 1783–1786 (1989).

K. Burton, "Simplex optimization," *Anal. Proc. (London)*, **26**(8), 285–288 (1989).

M. R. Cave and J. A. Forshaw, "Simplex optimization of response-time-limited systems. A study with a spectrophotometric determination of aluminum and mathematical test functions," *Anal. Chim. Acta*, **223**(2), 403–410 (1989).

A. Cullaj, "Simplex optimization of flame factors for the determination of calcium
by atomic absorption spectrometry," *Bul. Shkencave Nat.*, **43**(2), 41–45 (1989)
[Albanian, *Chem. Abstr.*, **112**(14), 131435z].

S. N. Deming, J. M. Palasota, J. Lee, and L. Sun, "Computer-assisted optimization
in high-performance liquid chromatographic method development," *J. Chromatogr.*, **485**, 15–25 (1989).

L. Ebdon, E. H. Evans, and N. W. Barnett, "Simplex optimization of experimental
conditions in inductively coupled plasma atomic emission spectrometry with
organic solvent introduction," *J. Anal. At. Spectrom.*, **4**(6), 505–508 (1989).

J. W. Elling, L. J. De Koning, F. A. Pinkse, N. M. M. Nibbering, M. M. Nico, and H.
C. Smit, "Computer-controlled simplex optimization on a Fourier-transform ion
cyclotron resonance mass spectrometer," *Anal. Chem.*, **61**(4), 330–334 (1989).

J. E. Haky, D. A. Sherwood, and S. T. Brennan, "Simplex optimization of densitome-
ter parameters for maximum precision in quantitative thin-layer chromatography," *J. Liquid Chromatogr.*, **12**(6), 907–917 (1989).

A. G. Howard and I. A. Boenicke, "An optimization function and its application to
the multidimensional thin-layer chromatography of protein amino acids," *Anal.
Chim. Acta*, **223**(2), 411–418 (1989).

P. C. Jurs, "Computer-controlled simplex optimization on a Fourier transform ion
cyclotron resonance mass spectrometer," *Chemtracts: Anal. Phys. Chem.*, **1**(4),
227–228 (1989).

J. Krupcik, D. Repka, T. Hevesi, E. Benicka, and J. Garaj, "On the use of sequential
simplex procedure for optimization of basic parameters influencing temperature
in LTPGC analysis of multicomponent samples," *Chromatographia*, **27**(7–8),
367–370 (1989).

E. P. Lankmayr, W. Wegscheider, and K. W. Budna, "Global optimization of HPLC
separations," *J. Liquid Chromatogr.*, **12**(1,2), 35–58 (1989).

I. Lavagnini, P. Pastore, and F. Magno, "Comparison of the simplex, Marquardt, and
extended and iterated Kalman filter procedures in the estimation of parameters
from voltammetric curves," *Anal. Chim. Acta*, **223**(1), 193–204 (1989).

R. E. Lenkinski, T. Allman, J. D. Scheiner, and S. N. Deming, "An automated iterative algorithm for the quantitative analysis of *in vivo* spectra based on the simplex optimization method," *Mag. Reson. Med.*, **10**, 338–348 (1989).

Y. G. Olesov, A. N. Rubtsov, and S. G. Mirontsov, "Optimization of the processes of dehydridation and nitridation of titanium in a pseudoliquefied state," *Izv. Akad. Nauk SSSR, Neorg. Mater.*, **25**(3), 519–521 (1989) [Russ., *Chem. Abstr.*, **111**(2), 16720f].

L. Peichang and H. Hongxin, "An intelligent search method for HPLC optimization," *J. Chromatogr. Sci.*, **27**, 690–697 (1989).

G. R. Phillips and E. M. Eyring, "Exchange of comments on the simplex algorithm culminating in quadratic convergence and error estimation," *Anal. Chem.*, **61**, 1786–1787 (1989).

T. Provder, "Cure characterization in product research and development," *J. Coatings Technol.*, **61**(770), 32–50 (1989).

S. Sangsila, G. Labinaz, J. S. Poland, and G. W. van Loon, "An experiment on sequential simplex optimization of an atomic absorption analysis procedure," *J. Chem. Educ.*, **66**(4), 351–353 (1989).

J. P. Schmit and A. Chauvette, "Simplex approach to the optimization of the ion optics bias potentials of an inductively coupled plasma mass spectrometer," *J. Anal. At. Spectrom.*, **4**(8), 755–759 (1989).

D. M. Sobolev, M. I. Il'in, S. V. Valgin, and Yu. V. Sharikov, "Use of the simplex framework method in quantitative gas-chromatographic analysis," *Zavod. Lab.*, **55**(5), 98–100 (1989) [Russ., *Chem. Abstr.*, **111**(24), 224221w].

D. M. Sobolev, Yu. V. Sharikov, and G. B. Selekhova, "Use of regression models for the calculation of chromatograms with overlapping peaks," *Zh. Anal. Khim.*, **44**(7), 1266–1273 (1989) [Russ., *Chem. Abstr.*, **111**(26), 247274e].

C. J. Thoennes and V. E. McCurdy, "Evaluation of a rapidly disintegrating, moisture resistant lacquer film coating," *Drug Dev. Ind. Pharm.*, **15**, 165–185 (1989).

H. R. Wilk and S. D. Brown, "Modifications to the simplex-optimized adaptive Kalman filter," *Anal. Chim. Acta*, **225**, 37–52 (1989).

P. J. Van Niekerk and R. A. Hasty, "Optimization of a technique for the estimation of the composition of edible oil blends," *Anal. Chim. Acta*, **223**(1), 237–246 (1989).

P. Zhang and Y. Ren, "Flexible tolerance simplex method and its application to multicomponent spectrophotometric determinations," *Anal. Chim. Acta*, **222**(2), 323–333 (1989).

1990:

J. A. Crow and J. P. Foley, "Optimization of separations in supercritical fluid chromatography using a modified simplex algorithm and short capillary columns," *Anal. Chem.*, **62**, 378–386 (1990).

J. W. Dolan and L. R. Snyder, "Integration of computer-aided method development techniques in LC," *J. Chromatogr. Sci.*, **28**, 379–384 (1990).

J. L. Glajch and L. R. Snyder, Eds., *Computer-Assisted Method Development for High-Performance Liquid Chromatography*, Elsevier, Amsterdam, 1990.

S. R. Goode, J. J. Gemmill, and B. E. Watt, "Determination of deuterium by gas chromatography with a microwave-induced plasma emission detector," *J. Anal. At. Spectrom.*, **5**, 483–486 (1990).

Y. Hu, J. Smeyers-Verbeke, and D. L. Massart, "An algorithm for fuzzy linear calibration," *Chemom. Intell. Lab. Sys.*, **8**, 143–155 (1990).

J. Sneddon, "Simplex optimization in atomic spectroscopy," *Spectroscopy*, **5**(7), 33–36 (1990).

E. Morgan, K. W. Burton, and G. Nickless, "Optimization using the modified simplex method," *Chemom. Intell. Lab. Sys.*, **7**, 209–222 (1990).

E. Morgan and K. W. Burton, "Optimization using the super-modified simplex method," *Chemom. Intell. Lab. Sys.*, **8**, 97–107 (1990).

Appendix A

Worksheets for Simplex Optimization

This appendix contains worksheets for use with two-, three-, and four-factor optimizations using fixed- or variable-size simplexes.

Simplex No. ___ → ___	Factor		Response	Rank	Vertex Number	Times Retained
	X_1	X_2				
Coordinates of				B		
retained vertexes				N		
Σ						
$\bar{P} = \Sigma/k$						
W				W		
$(\bar{P} - W)$						
$R = \bar{P} + (\bar{P} - W)$				R		0

Worksheet A.1 Fixed-size, two-factor worksheet.

Simplex No. ___ → ___

	Factor			Response	Rank	Vertex Number	Times Retained
	X_1	X_2	X_3				
Coordinates of					B		
					...		
retained vertexes					N		
Σ							
$\bar{P} = \Sigma/k$							
W					W		
$(\bar{P} - W)$							
$R = \bar{P} + (\bar{P} - W)$					R		0

Worksheet A.2 Fixed-size, three-factor worksheet.

Simplex No. ___ → ___

	Factor				Response	Rank	Vertex Number	Times Retained
	X_1	X_2	X_3	X_4				
Coordinates of						B		
						...		
retained vertexes						...		
						N		
Σ								
$\bar{P} = \Sigma/k$								
W						W		
$(\bar{P} - W)$								
$R = \bar{P} + (\bar{P} - W)$						R		0

Worksheet A.3 Fixed-size, four-factor worksheet.

Simplex No. ___ → ___	Factor		Response	Rank	Vertex Number	Times Retained
	X_1	X_2				
Coordinates of				B		
retained vertexes				N		
Σ						
$\bar{P} = \Sigma/k$						
W				W		
$(\bar{P} - W)$						
$R = \bar{P} + (\bar{P} - W)$				R		0
$(\bar{P} - W)/2$						
$C_W = \bar{P} - (\bar{P} - W)/2$				C_W		0
$C_r = \bar{P} + (\bar{P} - W)/2$				C_r		0
$E = R + (\bar{P} - W)$				E		0

Worksheet A.4   Variable-size, two-factor worksheet.

Simplex No. ___ → ___	Factor			Response	Rank	Vertex Number	Times Retained
	X_1	X_2	X_3				
Coordinates of					B		
					...		
retained vertexes					N		
Σ							
$\bar{P} = \Sigma/k$							
W					W		
$(\bar{P} - W)$							
$R = \bar{P} + (\bar{P} - W)$					R		0
$(\bar{P} - W)/2$							
$C_W = \bar{P} - (\bar{P} - W)/2$					C_W		0
$C_r = \bar{P} + (\bar{P} - W)/2$					C_r		0
$E = R + (\bar{P} - W)$					E		0

Worksheet A.5   Variable-size, three-factor worksheet.

Simplex No. ___ → ___	Factor				Response	Rank	Vertex Number	Times Retained
	x_1	x_2	x_3	x_4				
						B		
Coordinates of						...		
retained vertexes						...		
						N		
Σ								
$\bar{P} = \Sigma/k$								
W						W		
$(\bar{P} - W)$								
$R = \bar{P} + (\bar{P} - W)$						R		0
$(\bar{P} - W)/2$								
$C_W = \bar{P} - (\bar{P} - W)/2$						C_W		0
$C_r = \bar{P} + (\bar{P} - W)/2$						C_r		0
$E = R + (\bar{P} - W)$						E		0

Worksheet A.6 Variable-size, four-factor worksheet.

Appendix B

Glossary

accuracy (noun). The accuracy of a measurement signifies the closeness with which the measurement approaches the true value. It normally refers to the difference (error or bias) between the mean, $\bar{x}$, of the set of results and the true or correct value for the quantity measured, μ. Accuracy may also refer to the difference between an individual value x_i and μ. The relative accuracy of the mean is given by $(\bar{x} - \mu)/\mu$, and the percentage accuracy by $100(\bar{x} - \mu)/\mu$.

algorithm (noun). A prescribed set of well-defined rules or procedures for the solution of a problem in discrete steps.

analysis of variance (noun). Abbreviated ANOVA. A statistical technique allowing the total variance of a set of experiments to be separated into various contributions and providing tests of significance for those contributions.

ANOVA See "analysis of variance."

array (noun). A set or list of elements arranged in a sequence.

bias (noun). In general, an effect that deprives a statistical result of representativeness by systematically distorting it (as distinct from a nonsystematic or random effect that may distort on any one occasion but balances out on the average). See "accuracy," "systematic error."

bimodal (adjective). Having two local optima.

boundary (noun). See "constraint."

central composite design (noun). Also called a "star-square" design. A composite design obtained by superimposing a two-level factorial or fractional factorial design with a star design. See "composite designs."

centroid (noun). Center of mass.

composite designs (noun). Designs developed by Box and Wilson [*J. R. Stat. Soc.,* *B*, **13**, 1 (1951)] for determining the best fitting equation of second degree to represent a response surface without undue expenditure of experiments. Usually two-level factorial or fractional factorial designs allowing estimates of all linear and two-factor interaction effects augmented by extra points from a star design allowing quadratic effects to be determined. [G. E. P. Box, *Biometrics*, **10**, 16 (1954).]

confidence limits (or intervals) (noun). A measure of precision of an estimated parameter or response; the limits around the measured parameter or response within which the mean value for an infinite number of measurements (i.e., the true value) can be expected to be found with the stated level of probability. Confidence limits for the mean based on independent normally distributed measurements are given by

$$\text{confidence limit} = \bar{x} \pm ts/\sqrt{n}$$

where s is the standard deviation and t is the tabular or critical value of Student's t at the stated confidence level and number of degrees of freedom.

constraint (noun). Any limit imposed on a system (by the experimenter or the experiment) requiring some of the system variables to be restricted to specified values. An equality constraint

$$g(x_1, x_2, \ldots, x_n) = 0$$

restricts some functional relationship of system variables to be constant (e.g., $x_1 + x_2 = $ constant). An inequality constraint

$$h(x_1, x_2, \ldots, x_n) \le 0,$$

restricts some functional relationship of system variables to a specified range (e.g., $x_{\text{lower limit},i} \le x_i \le x_{\text{upper limit}, i}$). In either case, the region of search for an optimum is restricted.

contour map (noun). A graphic representation of a response surface giving curves of constant response.

decision (noun). A choice made between alternative courses of action.

degrees of freedom (noun). (1) In general, the difference between the number of observations and the number of parameters to be estimated. (2) The parameter ν (nu) of the t-distribution is its degrees of freedom. The parameters ν_1 and ν_2 of the F-distribution are the numerator and denominator degrees of freedom.

edge (noun). A line joining any two vertexes in a simplex.

effect (noun). In experimental design, the "effect" is a quantity (usually a model parameter to be estimated) that represents the change in response produced by a change in level of one or more of the factors.

efficiency (noun). By a more "efficient" experimental design is meant one providing more knowledge or information and a higher degree of precision in the same number of experiments as the design to which it is compared.

empirical feedback (noun). Routine, automatic feedback. The type of feedback where a particular response pattern leads, more or less automatically, to a particular action.

estimate (noun). When a value of a parameter is predicted based on information contained in a particular sample, the resulting predicted value is called an estimate. The estimator of a parameter β is denoted b. Two desirable properties of estimates are (1) that they be unbiased, and (2) that they have minimum variance.

EVOP (Evolutionary Operation) (noun). EVOP is the acronym for *E*Volution *OP*eration: multifactor response surface designs (factorial designs) combined with regression techniques to estimate the direction of steepest ascent and thereby to eventually locate the optimum region of a response surface. Pioneered by G. E. P. Box in the early 1950s, and now an established response surface method in industry.

experimental optimization (noun). The attainment of best response from a system by adjusting a set of experimental factors (where the response is subject to noise).

F (noun). The variance ratio, defined as the ratio of two independent estimates of variance:

$$F = s_1^2/s_2^2$$

The *F*-test is a statistical test concerning the equality of two population variances; the *F*-distribution arises as the sampling distribution for this ratio.

face (noun). A k-dimensional (hyper) surface bounding a simplex, where k is usually $< n$, the dimension of the factor space.

factor (noun). A variable under examination in an experiment as a possible cause of variation in the response of a system, e.g., in a "factorial" experiment. Note that some factors may be hidden – not under investigation and not under experimental control.

factor tolerance (noun). "The deviations from [the optimum values] that can be tolerated without causing more than a stated deterioration in the quality of results, thus ensuring a close enough approach to optimum values without waste of time, effort, and money on unnecessarily fine control" [A. L. Wilson, *Talanta*, **17**, 21 (1970)].

factor space (noun). For k factors, the region of experimentation available may be considered a k-dimensional "factor space"; i.e., that space consisting of all possible combinations of levels in the domains of all the factors.

factorial design (noun). A full (or complete) factorial design is a set of experiments conducted at all combinations of each factor (variable) and each level. For example, a 3 × 4 factorial design is a two-factor design in which one factor has three levels of experiments and the other factor has four (requiring experiments at 12 factor combinations). A 2^3 factorial would have three factors at two levels each, and require eight experiments at the vertexes of a cube.

factor space (noun). For k factors, the region of experimentation available may be considered a k-dimensional "factor space"; i.e., that space consisting of all possible combinations of levels in the domains of all the factors.

feasible region (noun). That region of factor space (to be searched for an optimum) in which the constraints are satisfied.

feedback (*noun*). The return to the input of a part of the output of a system to provide self-adaptive control.

fractional factorial design (*noun*). A subset of the full factorial design (omitting certain chosen levels) often employed to reduce the number of experiments when the number of factors is large. Fractional replication, as it is called, requires that the linear combination of observations used to estimate some of the effects (i.e., contrasts) be the same for the estimate of certain other effects. Factor effects that are estimated together in this way are said to be confounded and are aliases of one another. Often some effects may be expected on a priori grounds to be insignificant, with estimation of other effects that are real. Usually, treatments are chosen to estimate main effects and some of the lower order interactions; higher order interactions, although confounded with these effects, will not bias the results if truly insignificant. The result is higher efficiency.

global optimum (*noun*). That point on a response surface that is the overall optimum, the overall best (maximum of minimum).

hyper (*prefix*). That exists in a higher dimensional space, e.g., hypercube, hypersphere, hyperspace, hypersurface.

hypo (*prefix*). That exists in a lower dimensional space, e.g., hypocube, hyposphere, hypospace, hyposurface.

hypothesis (*noun*). A tentative theory or supposition that is provisionally adapted to explain certain facts and to guide in the investigation of others. When used in the statistical sense, it is an assumption about the form of a population or its parameters. Hypotheses are denoted by the letter "H" and are usually stated in an exact mathematical equation.

identity matrix (*noun*). A square matrix having diagonal elements all equal to unity and off-diagonal elements all equal to zero; denoted I.

infeasible region (*noun*). That region of factor space in which the constraints are not satisfied, and therefore not available to the experimenter in the search for the optimum.

information (*noun*). Data that reduce uncertainty.

input (*noun*). A factor.

interaction (*noun*). In general, a joint effect of two or more variables. In factorial experiments it is a measure of the extent to which a change in response produced by changes in the levels of one or more factors depends on the levels of one or more other factors.

iteration (*noun*). The process of carrying out a specified set of operations.

iterative loop (*noun*). A means by which a set of specified operations is performed repeatedly.

least-squares, method of (*noun*). "For linear least squares, ... Gauss (1821) showed that if we have n observations $y_1, y_2, \ldots, y_n$ and *if* an appropriate model for the uth observation is

$$y_u = \beta_0 + \beta_1 x_{1u} + \beta_2 x_{2u} + \cdots + \beta_k x_{ku} + r_u$$

where the β's are unknown parameters, the x's known constants, and the r's random variables uncorrelated and having the same variance and zero expectation, then estimates $b_0, b_1, \ldots, b_k$ of the β's obtained by minimizing $\Sigma(y - \hat{y})^2$ with $\hat{y} = b_0 x_0 + b_1 x_1 + b_2 x_2 + \cdots + b_k x_k$ are unbiased and have smallest variance among all linear unbiased estimates" [G. E. P. Box, *Technometrics*, **8**(4), 625 (1966)]. The method of least-squares gives a powerful tool for curve-fitting and approximating a response surface.

level (*noun*). The amount, value, or intensity at which a factor in an experimental design is held fixed for a particular experiment. For example, a design might specify three levels of temperature at which experiments are to be done: $91°$, $96°$, and $101°$C. A level may be quantitatively measured (as for a continuous variable) or qualitatively measured (as for a discrete variable).

local optimum (*noun*). That point on a response surface that is only an optimum within some local area of search. It is usually very difficult to determine if a local optimum is also the global optimum unless all the possible local optima are exhaustively evaluated.

loop (*verb*). To execute a sequence of instructions repeatedly until a terminal condition prevails. See "iteration," "iterative loop."

main effect (*noun*). In factorial experiments, the main effect of a factor is the linear effect produced by averaging the changes in response over all possible combinations of levels of the other factors, i.e., the average change in response produced by changing the level of a single factor.

matrix (*noun*). A rectangular array of elements. A vector is a one-dimensional matrix. A table is (usually) a two-dimensional matrix. If a matrix is denoted A, A' denotes the transpose of the matrix A, obtained by exchanging the elements of the rows with the elements of the columns. A^{-1} denotes the inverse of the square matrix A, where $A\,A^{-1} = A^{-1}A = I$, the identity matrix.

maximum (*noun*). The greatest value attainable; that point on a response surface yielding the greatest response.

mean (*noun*). Also known as the arithmetic mean or average, the mean of n numbers is their sum divided by n:

$$\bar{x} = (x_1 + x_2 + \cdots + x_n)\,/n/$$

This is also the most widely used measure of the middle or center of a set of data. Population means are usually denoted by the greek letter μ (mu); sample means by $\bar{x}$.

mean deviation (*noun*). One of the early measures of precision not in favor today. The average of the absolute differences between each value and the average,

$$d_\mathrm{m} = (1/n) \sum_{i=1}^{n} |x_i - \bar{x}{\cdot}|$$

The mean deviation is largely replaced by the standard deviation. The mean deviation is not recommended as a measure of precision except when the set consists of only a few measurements.

measurement (noun). The assignment of numerical quantities to represent properties.

minimum (noun). The smallest value attainable; that point on a response surface yielding the smallest response.

model (noun). An expression of a causal situation attempting to describe aspects of a process (or phenomenon) that has generated observed data. In a statistical sense, "model" usually refers to a mathematical statement (using the model parameters and other symbols in an equation) used in studying the results of an experiment or predicting the behavior of future repetitions of the experiment. A model is an equation relating the responses to the factors and thereby approximating the transfer function of the real system.

multimodal (adjective). Having more than one optima.

noise (noun). In information theory, noise refers to random disturbance superimposed on a signal.

nonfeasible region (noun). See "infeasible region."

normal distribution (noun). The Gaussian, the Laplacean, the Gauss–Laplace distribution, the normal curve of error, or simply the "bell" curve is the continuous frequency distribution of infinite range given by

$$f = [1/(\sigma\sqrt{2\pi})]\exp\{-\tfrac{1}{2}[(x - \mu)/\sigma]^2\}, \quad -\infty \leq x \leq \infty,$$

where f is the normalized frequency, μ is the mean, and σ is the standard deviation. First discovered in 1756 by De Moivre as the limiting form of the binomial distribution.

null hypothesis (noun). A hypothesis, literally, of no differences, i.e., a tentative assumption that the treatment will have no effect, or that there is no difference between two estimates.

numerical optimization (noun). The attainment of best response from a system, generally a deterministic system, by adjusting a set of parameters.

observational data (noun). Data acquired by merely observing a system passively, rather than by perturbing a system with experiments.

optimization (noun). The process by which the response of a system is improved to its best possible (maximized or minimized).

optimum (noun). Word coined by Leibniz (1710). The best. Plural is optima.

output (noun). A response.

parameter (noun). In statistics, a parameter is a numerical quantity that characterizes the distribution or population of a random variable and that could be calculated if the entire population were available. Examples are the parameters μ and σ, the mean and standard deviation of a population. In general, a parameter is any unknown quantity that may vary or be adjusted over a certain set of values. The βs in linear models are parameters.

point estimation (noun). The calculation of a single number that determines a point on a line and that provides an estimate of the parameter of interest. See "estimate."

population (*noun*). A set or collection of responses, usually quantitative, existing in fact or concept. The set may be infinite or finite in number.

precision (*noun*). The precision of a set of measurements signifies the closeness with which the set of measurements approaches the average of a long series of measurements under similar conditions; precision is concerned with the repeatability of measurements within a set. The sample standard deviation, *s*, becomes more reliable as a measure of precision as the sample size increases. The range is not recommended as a measure of precision except when the number of measurements in the set is small.

probabilistic models (*noun*). A mathematical model that contains one or more random components intended to explain the apparent random variability of the response for given values of the variables: e.g.,

$$y_{1i} = \beta_0 + \beta_1 x_{1i} + r_{1i}$$

where r_{1i} is the random component with a specified probability distribution, expected value, and variance.

qualitative factor (*noun*). A factor whose different levels are not susceptible to quantitative distinction; e.g., the choice of solvent for a chemical reaction in which only acetone, acetic acid, tetrahydrofuran, or benzene can be used.

random (*adjective*). Implies the process under consideration is in some sense probabilistic. See "stochastic process."

randomization (*noun*). In an experimental design the order of application of the treatments should be a random one. For example, if eight trials of a 2^3 factorial are to be done, they should be run in a random order. Randomization is simply allowing chance to determine the order of experiments. The effect of this is to provide "insurance" against possible systematic biases "that may or may not occur and that may or may not be serious if they do occur." Randomization should be done by some random process, e.g., flipping a coin or using a table of random numbers.

range (*noun*). The difference in magnitude between the smallest and largest measurements in a set. Range is not recommended as a measure of precision except when the number of measurements is small; if used, the number of measurements should be reported.

rank (*noun*). Relative position in a set of results.

rank (*verb*). To determine the relative position of a result or set of results.

replicate (*noun*). A design in which each treatment combination or experiment is applied once is a single replicate of the experimental design. If each treatment is applied *m* times, the design is said to be replicated *m* replicates.

replicate (*verb*). In experimental design, the performance of an experiment more than once to obtain more information (degrees of freedom) for estimating experimental error and to obtain estimates of factor effects with improved precision.

response (*noun*). A measured output from a system.

response surface (*noun*). A graph of the response of a system of *k* factors drawn as a surface in $(k + 1)$-dimensional space.

robust (adjective).   Showing strength; performing well. Used in reference to statistical techniques when they perform well, especially in the presence of experimental error.

saddle point (noun).   A col or minimax point on a response surface.

sample (noun).   A subset or part of a population containing information, albeit incomplete, concerning the population.

sample (verb).   To obtain data concerning a subset of a population; to take a sample.

scedasticity (noun).   Dispersion, especially measured by variance. A distribution is said to be "homoscedastic" in a variable if the variance is the same for all fixed values of that variable; if not, it is "heteroscedastic."

scientific feedback (noun).   Feedback in which experimental results interact with technical knowledge to produce actions that could not be taken on a purely automatic basis.

self-adaptation (noun).   The ability of a system to modify itself in response to changes in its environment.

sequential simplex method (noun).   A highly efficient, multifactor empirical feedback search strategy for optimization that was introduced by Spendley, Hext, and Himsworth, *Technometrics*, **4**, 441 (1962), and modified by Nelder and Mead, *Comput. J.*, **7**, 308 (1965). It is a hill-climbing algorithm that moves a pattern of $(k + 1)$ experimental points (for k factors) away from regions of worse response toward convergence on an optimum in the response surface.

simplex (noun).   That geometric figure defined by a number of points ($k + 1$ vertexes) equal to one more than the number of dimensions of the factor space (k). A simplex in two dimensions is a triangle; in three dimensions, a tetrahedron; the series can be extended to higher dimensions but the simplexes are not easily visualized as geometric figures.

sort (verb).   To arrange a set of items in a new order according to a specified rule. See "rank."

standard deviation (noun).   A measure of precision of a variate; the population standard deviation, σ, is estimated by the sample standard deviation,

$$s = \sqrt{[1/(n-1)] \sum_{i=1}^{n} (x_i - \bar{x})^2}$$

where $\bar{x}$ is the mean of the set of n measurements x_i. The standard deviation has the same units as the measurement. It becomes a more reliable expression of precision as n becomes larger. See "variance," which is the standard deviation squared.

stationary point (noun).   That point in the factor space where the first partial derivatives of the response surface equation with respect to each factor are simultaneously equal to zero (not necessarily an optimum).

steepest ascent (noun).   That limited direction in factor space from a local region that will yield the highest improvement in the response.

stochastic (adjective).   Implying the presence of a random component. For opposite, see "deterministic."

stochastic process (*noun*).   Any process that can be described in terms of probabilities.

surface (*noun*).   See "response surface."

system (*noun*).   A set or arrangement of a set of connected or related objects (elements) and the set of relationships among those objects. Most interesting systems are dynamic: information and energy are exchanged between their parts; they change with time.

systematic error (*noun*).   Bias. As opposed to random error, an error that is biased, i.e., has a distribution with mean not equal to zero. See "bias," "accuracy."

systems theory (*noun*).   Delineation of systems in terms of functional parts and boundaries; definition of inputs and outputs.

t-distribution (*noun*).   A distribution used for inferences concerning the mean(s) of normal distributions whose variances are unknown; derivation due to W. S. Gossett who used the pen name "Student."

tangent (*noun*).   A line or (hyper)plane that touches a curve or (hyper)surface at a single point.

tetrahedron (*noun*).   A polyhedron with four triangular faces and four vertexes. A pyramid on a triangular base. A three-dimensional simplex.

transformation of scale (*noun*).   A change of label and associated numerical change; e.g., 5 grams → 5000 milligrams.

treatments (*noun*).   The different procedures whose effects are to be measured and compared; experiments.

unimodal (*noun*).   Having only a single optimum.

variable (*noun*).   Any quantity that varies; a quantity that may take any one of a specified set of values. An independent variable is usually a variable that is unaffected by changes in the levels of other variables. A dependent variable is affected by changes in the levels of other variables.

variance (*noun*).   The average of the square of the deviations of the measurements about their mean. The variance of a population is denoted σ^2, while the variance of a sample is denoted s^2. See "precision," "standard deviation."

variance, unbiased estimate of (*noun*).   Computed from a sample of n measurements:

$$s = [1/(n-1)] \sum_{i=1}^{n} (x_i - \bar{x})^2$$

referred to as the sample variance.

vertex (*noun*).   In a simplex, a vertex refers to any point opposite a (hyper)face of the simplex where for k factors, k edges of the simplex intersect. In general, a vertex is the intersection of two or more lines or curves.

Index